AF361905

The Drought Risk Analysis, Forecasting, and Assessment under Climate Change

The Drought Risk Analysis, Forecasting, and Assessment under Climate Change

Editor

Tae-Woong Kim

MDPI • Basel • Beijing • Wuhan • Barcelona • Belgrade • Manchester • Tokyo • Cluj • Tianjin

Editor
Tae-Woong Kim
Palacky University
Czech Republic

Editorial Office
MDPI
St. Alban-Anlage 66
4052 Basel, Switzerland

This is a reprint of articles from the Special Issue published online in the open access journal *Water* (ISSN 2073-4441) (available at: https://www.mdpi.com/journal/water/special_issues/drought_risk).

For citation purposes, cite each article independently as indicated on the article page online and as indicated below:

LastName, A.A.; LastName, B.B.; LastName, C.C. Article Title. *Journal Name* **Year**, *Article Number*, Page Range.

ISBN 978-3-03936-806-8 (Hbk)
ISBN 978-3-03936-807-5 (PDF)

Contents

About the Editor

Tae-Woong Kim, Ph.D., is a professor in the Department of Civil and Environmental Engineering at Hanyang University (ERICA). His research on water resources engineering centers on water resources planning, risk assessment, climate change, prediction and forecasting, and machine learning. His published research can be found in a variety of journals, including the *Journal of Hydrology, Journal of Irrigation and Drainage Engineering, Stochastic Environmental Research and Risk Assessment, Terrestrial Atmospheric and Oceanic Sciences, Journal of Hydro-environment Research, Water Resources Management, KSCE Journal of Civil Engineering, and Water*.

Editorial

Drought Risk Analysis, Forecasting and Assessment under Climate Change

Tae-Woong Kim [1,*] and Muhammad Jehanzaib [2]

[1] Department of Civil and Environmental Engineering, Hanyang University, Ansan 15588, Korea
[2] Department of Civil and Environmental Engineering, Hanyang University, Seoul 04763, Korea; jehanzaib7@hanyang.ac.kr
*   Correspondence: twkim72@hanyang.ac.kr

Received: 21 June 2020; Accepted: 26 June 2020; Published: 29 June 2020

Abstract: Climate change is undoubtedly one of the world's biggest challenges in the 21st century. Drought risk analysis, forecasting and assessment are facing rapid expansion, not only from theoretical but also practical points of view. Accurate monitoring, forecasting and comprehensive assessments are of the utmost importance for reliable drought-related decision-making. The framework of drought risk analysis provides a unified and coherent approach to solving inference and decision-making problems under uncertainty due to climate change, such as hydro-meteorological modeling, drought frequency estimation, hybrid models of forecasting and water resource management. This Special Issue will provide researchers with a summary of the latest drought research developments in order to identify and understand the profound impacts of climate change on drought risks and water resources. The ten peer-reviewed articles collected in this Special Issue present novel drought monitoring and forecasting approaches, unique methods for drought risk estimation and creative frameworks for environmental change assessment. These articles will serve as valuable references for future drought-related disaster mitigations, climate change interconnections and food productivity impacts.

Keywords: climate change; drought risk; assessment; forecasting

---

1. Introduction

Drought is a complex natural catastrophe that has plagued civilization throughout history. It usually develops when different hydro-meteorological variables encounter drier than normal conditions. Most parts of the world, even wet and humid regions, suffer from drought, though arid areas are more susceptible because their moisture levels rely on only a few critical rainfall events [1]. Drought can be categorized into several types depending on its impact on the hydrological cycle [2]. Lack of precipitation for an extended period, i.e., several months to years, leads to meteorological drought [3]. Meteorological drought inevitably spreads through the hydrological cycle for a prolonged duration. This may cause crop yield reductions due to reduced soil moisture and is known as agricultural drought [4,5]. Extended meteorological drought may cause a streamflow shortage, called hydrological drought [6].

Drought is monitored using drought indices, which measure deviations from normal local conditions based on historical distributions [7]. Drought indices developed to quantify meteorological, agricultural and hydrological drought include the Rainfall Anomaly Index (RAI) [8], Palmer Drought Severity Index (PDSI) [9], Standardized Precipitation Index (SPI) [10], Reconnaissance Drought Index (RDI) [11], Standardized Precipitation Evapotranspiration Index (SPEI) [12], Crop Moisture Index (CMI) [13], Soil Moisture Drought Index (SMDI) [14] and Standardized Runoff Index (SRI) [15]. Standardized drought indices are widely used due to their versatility and simplicity over various time scales [6], and most only employ a single input indicator for drought monitoring. Single indicator-based drought indices do not fully reflect drought information and may lead to unreliable results [16]. They are also based on the presumption of stationarity, which is not a valid assumption under changing climate conditions; a more general and robust drought monitoring system is required.

Over the last century, the global climate and the environment have changed remarkably; global warming, which contributes to increased water circulation, has caused severe natural disasters including floods and droughts [17]. In most parts of the world, drought risk has increased significantly since 1970 due to the rise in evapotranspiration without any precipitation enhancement [7]. In East Asia, China is frequently affected by drought due to significant precipitation and temperature variations [18–20]. The Ministry of Water Resources of China [21] reported extreme droughts occurring at an average rate of every two years from 1990 to 2007. The associated crop productivity loss was estimated at almost 39.2 billion kg per year, about 1.47 percent of the gross domestic product. The Australian Agriculture and Resource Economics Bureau reported that the 2006 drought reduced national winter cereal crops by 36 percent and cost AUD \$3.5 billion, leaving numerous farmers in fiscal crisis [22]. Therefore, developing schemes for examining climate change-induced drought impacts on crop productivity is crucial for sustaining global agriculture.

Global temperature increases and rainfall pattern variations are evidence that drought frequency and severity are greatly affected by climate change [23]. Blenkinsop and Fowler [24] assessed climate change impacts on drought characteristics and reported that short-term summer droughts are projected to increase while long-term droughts will become less severe. Sheffield and Wood [25] used soil moisture data to examine shifts in drought incidence by combining multiple General Circulation Models (GCMs) and multi-scenarios. Extensive studies [26–30] focus on univariate and bivariate frequency analyses for drought risk assessment due to climate change. Anthropogenic activities, in addition to climate change, are also a major factor affecting drought phenomena [31]. Urbanization, variation in land use/land cover and industrialization can influence hydrological processes and exert environmental impacts, with substantial implications on water resources and, ultimately, hydrological drought [32]. Many studies have quantified the influence of climate change and human activity on streamflow, but studies on how they affect hydrological drought are very rare [33–35]. All of these investigations utilized conventional drought indices for drought monitoring and are based on stationarity presumptions; this is not valid for varying environmental conditions.

This Special Issue provides a platform for researchers to fill these gaps with their experience and expertise. Figure 1 shows the climate-influenced relationships between hydrological cycle variables and drought types. This Special Issue covers (1) robust index development for effective drought monitoring; (2) risk analysis framework development and early warning systems; (3) impact investigations on crop productivity; (4) environmental change impact analyses.

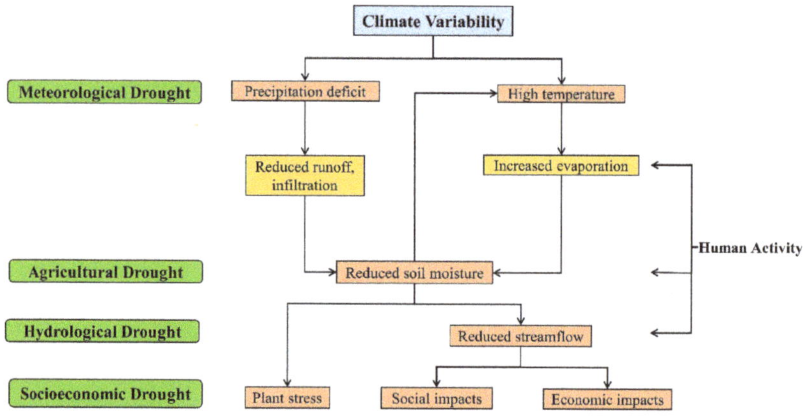

Figure 1. Climate-influenced relationships between hydrological cycle variables and drought types.

2. Special Issue Overview

This Special Issue includes 10 peer-reviewed articles covering a wide range of research topics related to drought monitoring, drought forecasting and drought risk analysis in a changing climate. Specific issues include the development of a modified composite drought index (MCDI) and a non-stationary joint drought management index (JDMI) [36,37], climate change influences on drought patterns and crop yields [38,39], meteorological and hydrological drought risk under future climate change predictions [40,41], extreme drought assessment and its relationship with the Indian Ocean dipole (IOD) mode [42], severe drought prediction using atmospheric teleconnection patterns (ATPs) [43], drought forecasting using stochastic models [44] and hydrological drought risk estimations based on changing climate conditions and human activities [45]. These studies use statistical approaches, field measurements and mathematical methodologies.

Chen et al. [36] developed a new multivariate drought index based on multiple input variables: precipitation, temperature, evaporation and surface water content. They compared the modified composite drought index (MCDI) to the meteorological drought composite index (CI), an existing multivariate drought index, and found a high correlation with drought events in China's Hubei Province. MCDI's drought monitoring accuracy is also compatible with historical drought records. MCDI is a reliable drought monitoring index that represents integrated meteorological and agricultural drought knowledge. In another study, Yu et al. [37] formulated a non-stationary joint drought management index (JDMI) for hydrological drought risk assessments. They formulated a bivariate time-varying copula model using generalized additive models for location, scale and shape (GAMLSS) and determined future low-flow drought using simulated data for South Korea's Soyang River basin. From this, they calculated water supply performance indexes, considering reliability and vulnerability. The outcomes suggest that time-varying models are more suitable for drought modeling under changing environmental conditions. The non-stationary JDMI may serve as a significant model for drought monitoring, planning and mitigation.

Guna et al. [38] examined the influence of climate variation on maize crop productivity during the growing season in Songliao Plain, China. They applied the Mann–Kendall mutation test to determine temperature and precipitation trends and used the standardized precipitation evapotranspiration index (SPEI) to examine drought characteristics. They estimated the relationship between meteorological drivers, drought indices and maize productivity and determined that the correlation between climate yield and temperature is negative, while the correlation between climate yield and precipitation and drought is positive. Using future global warming scenarios (1.5 °C and the 2.0 °C), the results show a decrease in predicted maize productivity from −7.7% to −15.9% using multiple regressions and −12.2% to 21.8% using one-variable regressions. These findings explain potential future security. Similarly, Qutbudin et al. [39] investigated seasonal drought pattern variability using the Mann–Kendall test and SPEI in Afghanistan. Their results indicate increasing Afghan drought severity and frequency. A 0.14 °C/decade temperature increase and rainfall decrease are the primary factors influencing the results, which occur during the rice, corn and wheat growing season in Northwest and Southwest Afghanistan. This methodology can be applied elsewhere and utilized in adaptation and mitigation policy development.

Kwon and Sung [41] examined future drought changes using HadGEM2-AO projections in South Korea. They projected future drought based on baseline climatic reference precipitation data and quantitatively assessed changes in future drought's severity and frequency. Their results, which are distinctly different from previous studies due to existing methodology modification, suggest that future drought will be weaker and less frequent due to a rise in precipitation. In future climates, mild drought will occur more frequently, but drought frequency based on baseline climate will decrease. In order to develop better coping strategies for future drought, knowledge about a region's baseline and future climates is essential. To better understand climate variation influences on drought, Kim et al. [40] quantified the future hydrologic risk of extreme drought in South Korea using climate change scenario representative concentration pathway (RCP) 8.5. They adopted the threshold level method to identify drought events and extract drought characteristics. They used bivariate frequency analysis to determine return periods, taking into account drought duration and drought severity. They calculated that the extreme drought hazard median will be higher in the future than the baseline period's maximum drought. These results will help establish drought risk-based quantitative design standards for water resource systems.

Global rainfall patterns have changed due to climate change, resulting in extreme natural disasters including drought. Yeh and Hsu [44] proposed an early warning system for drought forecasting using stochastic, autoregressive integrated moving average (ARIMA) models based on the standardized precipitation index (SPI) in Southern Taiwan. The ARIMA model yielded determination coefficients (R2) of more than 80% at each station, with sufficiently low error indicators. This suggests that the ARIMA model is a powerful tool for drought forecasting. Apart from climate change, anthropogenic activities also influence hydrological drought risk. Zhang et al. [45] quantified climate change and anthropogenic activity impacts on the hydrological drought risk for China's Kuye River basin (KRB). They found that the KRB's annual runoff pattern changed significantly after 1979, and drought characteristics (duration and severity) have been considerably worse in more recent times. The quantitative assessment reflects that human activities lead to an increased regional drought risk, and the modeling results can be utilized to plan and control sustainable water supplies.

Two studies improved extreme drought predictability by identifying the relationship between the Indian Ocean dipole (IOD) and atmospheric teleconnection patterns (ATPs). Gao et al. [42] examined extreme droughts in the Indochina peninsula (ICP) and its IOD relationship. They mimicked the drought-sensitive area and various IOD evolution patterns via statistical simulations. The results showed a reduction in extreme drought frequency throughout Vietnam and Southwestern China. In contrast, they saw a drought frequency increase in Cambodia, Central Laos, and along the coastline adjacent to the Myanmar Sea. Gao et al. [43] identified the relationship between extreme droughts and ATPs by selecting a core drought region (CDR) based on historical drought analysis. They chose four principal components (PCs) based on eight teleconnection variances. The extreme spring drought (ESD) predictions showed that the neural network's predictive performance was superior to the Poisson regression. These studies will be helpful in improving extreme drought diagnostic methods.

3. Conclusions

This Special Issue's ten articles advance our understanding of drought's complex phenomena and its interaction with climate change and human activity. The newly proposed drought indices [36,37] will serve as effective tools for drought monitoring under changing environmental conditions. The indices can incorporate multiple inputs for drought calculation, which is more realistic than traditional methods. Two articles [38,39] investigated the effects of climate change-induced drought on various types of crop productivity. These studies showed that temperature increases result in decreases in crop productivity; this is linked to food security. The next two articles [40,41] examined meteorological and hydrological drought risks under future climate change scenarios. The methodologies presented in these two articles are helpful to cope with future natural disasters. Two papers [42,43] improve drought predictability by identifying the relationships between drought and the Indian Ocean dipole (IOD) and atmospheric teleconnection patterns (ATPs). Another article [44] proposed a drought early warning system using a stochastic, autoregressive, integrated moving average (ARIMA) model; this study is very beneficial because advance knowledge is required for effective management.

These articles cover a wide geographic range, across China [36,38,43,45], Taiwan [44], South Korea [37,40,41] and the Indo–China peninsula [42], which covers many contrasting climatic conditions. Hence, their results have global implications: the data, analysis/modeling, methodologies and conclusions lay a solid foundation for enhancing our scientific knowledge of drought's complex mechanisms and relationships to varying environmental conditions.

Author Contributions: Conceptualization, T.-W.K.; writing—original draft preparation, M.J.; writing—review and editing, T.-W.K. All authors have read and agreed to the published version of the manuscript.

Funding: This research received no external funding.

Acknowledgments: This special issue is partially supported by the Korea Environmental Industry & Technology Institute (KEITI) grants funded by the Korea Ministry of Environment (83070 and 79616). The authors highly appreciate the editor and anonymous reviewers for their invaluable comments as well as the publication team of Water for their tireless efforts.

Conflicts of Interest: The authors declare no conflict of interest.

References

1. Sun, Y.; Solomon, S.; Dai, A.; Portmann, R.W. How often does it rain? *J. Clim.* **2006**, *19*, 916–934. [CrossRef]
2. Heim, R.R., Jr. A review of twentieth-century drought indices used in the United States. *Bull. Am. Meteorol. Soc.* **2002**, *83*, 1149–1166. [CrossRef]
3. Wilhite, D.A. *Drought as a Natural Hazard: Concepts and Definitions*; Routledge: London, UK, 2000.
4. Keyantash, J.; Dracup, J.A. The quantification of drought: An evaluation of drought indices. *Bull. Am. Meteorol. Soc.* **2002**, *83*, 1167–1180. [CrossRef]
5. Jehanzaib, M.; Kim, T.W. Exploring the influence of climate change-induced drought propagation on wetlands. *Ecol. Eng.* **2020**, *149*, 105799. [CrossRef]
6. Jehanzaib, M.; Sattar, M.N.; Lee, J.H.; Kim, T.W. Investigating effect of climate change on drought propagation from meteorological to hydrological drought using multi-model ensemble projections. *Stoch. Environ. Res. Risk Assess.* **2020**, *34*, 7–21. [CrossRef]
7. Dai, A. Drought under global warming: A review. *Wiley Interdiscip. Rev. Clim. Chang.* **2011**, *2*, 45–65. [CrossRef]
8. Van-Rooy, M.P. A rainfall anomaly index (RAI) independent of time and space. *Notos* **1965**, *14*, 43–48.
9. Palmer, W.C. *Meteorologic Drought*; Weather Bureau, Research Paper No. 45; US Department of Commerce, Weather Bureau: Washington, DC, USA, 1965.
10. McKee, T.B.; Doesken, N.J.; Kleist, J. The relationship of drought frequency and duration to time scales. In Proceedings of the 8th Conference on Applied Climatology, Anaheim, CA, USA, 17–22 January 1993; Volume 17, pp. 179–183.
11. Tsakiris, G.; Vangelis, H.J.E.W. Establishing a drought index incorporating evapotranspiration. *Eur. Water* **2005**, *9*, 3–11.
12. Vicente-Serrano, S.M.; Beguería, S.; López-Moreno, J.I. A multiscalar drought index sensitive to global warming: The standardized precipitation evapotranspiration index. *J. Clim.* **2010**, *23*, 1696–1718. [CrossRef]
13. Palmer, W.C. Keeping track of crop moisture conditions, nationwide: The new crop moisture index. *Weatherwise* **1968**, *21*, 156–161. [CrossRef]
14. Hollinger, S.E.; Isard, S.A.; Welford, M.R. A New Soil Moisture Drought Index for Predicting Crop Yields. In *Preprints, Eighth Conference on Applied Climatology*; American Meteorological Society: Anaheim, CA, USA, 1993; pp. 187–190.
15. Shukla, S.; Wood, A.W. Use of a standardized runoff index for characterizing hydrologic drought. *Geophys. Res. Lett.* **2008**, *35*, L02405. [CrossRef]
16. Hao, Z.; AghaKouchak, A. Multivariate standardized drought index: A parametric multi-index model. *Adv. Water Resour.* **2013**, *57*, 12–18. [CrossRef]
17. Leng, G.; Tang, Q.; Rayburg, S. Climate change impacts on meteorological, agricultural and hydrological droughts in China. *Glob. Planet. Chang.* **2015**, *126*, 23–34. [CrossRef]
18. Ma, Z.; Fu, C. Interannual characteristics of the surface hydrological variables over the arid and semi-arid areas of northern China. *Glob. Planet. Chang.* **2003**, *37*, 189–200. [CrossRef]
19. Dai, A.; Trenberth, K.E.; Qian, T. A global dataset of Palmer Drought Severity Index for 1870–2002: Relationship with soil moisture and effects of surface warming. *J. Hydrometeorol.* **2004**, *5*, 1117–1130. [CrossRef]
20. Zou, X.; Zhai, P.; Zhang, Q. Variations in droughts over China: 1951–2003. *Geophys. Res. Lett.* **2005**, *32*. [CrossRef]
21. *MWRC China Water Resources Bulletin 2011*; China WaterPower Press: Beijing, China, 2011.

22. Wong, G.; Lambert, M.F.; Leonard, M.; Metcalfe, A.V. Drought analysis using trivariate copulas conditional on climatic states. *J. Hydrol. Eng.* **2010**, *15*, 129–141. [CrossRef]

23. IPCC, C.C. The physical science basis. In *Contribution of Working Group I to the Fourth Assessment Report of the Intergovernmental Panel on Climate Change*; Cambridge University Press: Cambridge, UK; New York, NY, USA, 2007.

24. Blenkinsop, S.; Fowler, H.J. Changes in drought frequency, severity and duration for the British Isles projected by the PRUDENCE regional climate models. *J. Hydrol.* **2007**, *342*, 50–71. [CrossRef]

25. Sheffield, J.; Wood, E.F. Projected changes in drought occurrence under future global warming from multi-model, multi-scenario, IPCC AR4 simulations. *Clim. Dyn.* **2008**, *31*, 79–105. [CrossRef]

26. Salas, J.D.; Fu, C.; Cancelliere, A.; Dustin, D.; Bode, D.; Pineda, A.; Vincent, E. Characterizing the severity and risk of drought in the Poudre River, Colorado. *J. Water Resour. Plann. Manag.* **2005**, *131*, 383–393. [CrossRef]

27. Santos, J.F.; Portela, M.M.; Pulido-Calvo, I. Regional frequency analysis of droughts in Portugal. *Water Resour. Manag.* **2011**, *25*, 3537. [CrossRef]

28. Yoo, J.; Kim, U.; Kim, T.W. Bivariate drought frequency curves and confidence intervals: A case study using monthly rainfall generation. *Stoch. Environ. Res. Risk Assess.* **2013**, *27*, 285–295. [CrossRef]

29. Mortuza, M.R.; Moges, E.; Demissie, Y.; Li, H.Y. Historical and future drought in Bangladesh using copula-based bivariate regional frequency analysis. *Theor. Appl. Climatol.* **2019**, *135*, 855–871. [CrossRef]

30. Jehanzaib, M.; Kim, J.E.; Park, J.Y.; Kim, T.W. Probabilistic Analysis of Drought Characteristics in Pakistan Using a Bivariate Copula Model. In Proceedings of the Korea Water Resources Association Conference, Yeosu, Korea, 30–31 May 2019; p. 151.

31. Jehanzaib, M.; Shah, S.A.; Yoo, J.; Kim, T.W. Investigating the impacts of climate change and human activities on hydrological drought using non-stationary approaches. *J. Hydrol.* **2020**, *588*, 125052. [CrossRef]

32. Sheffield, J.; Wood, E.F.; Roderick, M.L. Little change in global drought over the past 60 years. *Nature* **2012**, *491*, 435–438. [CrossRef]

33. Jiang, S.; Wang, M.; Ren, L.; Xu, C.Y.; Yuan, F.; Liu, Y.; Lu, Y.; Shen, H. A framework for quantifying the impacts of climate change and human activities on hydrological drought in a semiarid basin of Northern China. *Hydrol. Process.* **2019**, *33*, 1075–1088. [CrossRef]

34. Jehanzaib, M.; Shah, S.A.; Kwon, H.H.; Kim, T.W. Investigating the influence of natural events and anthropogenic activities on hydrological drought in South Korea. *Terr. Atmos. Ocean. Sci.* **2020**, *31*, 85–96. [CrossRef]

35. Zhang, D.; Zhang, Q.; Qiu, J.; Bai, P.; Liang, K.; Li, X. Intensification of hydrological drought due to human activity in the middle reaches of the Yangtze River, China. *Sci. Total Environ.* **2018**, *637*, 1432–1442. [CrossRef]

36. Chen, S.; Zhong, W.; Pan, S.; Xie, Q.; Kim, T.W. Comprehensive Drought Assessment Using a Modified Composite Drought index: A Case Study in Hubei Province, China. *Water* **2020**, *12*, 462. [CrossRef]

37. Yu, J.; Kim, T.W.; Park, D.H. Future hydrological drought risk assessment based on non-stationary joint drought management index. *Water* **2019**, *11*, 532. [CrossRef]

38. Guna, A.; Zhang, J.; Tong, S.; Bao, Y.; Han, A.; Li, K. Effect of Climate Change on Maize Yield in the Growing Season: A Case Study of the Songliao Plain Maize Belt. *Water* **2019**, *11*, 2108. [CrossRef]

39. Qutbudin, I.; Shiru, M.S.; Sharafati, A.; Ahmed, K.; Al-Ansari, N.; Yaseen, Z.M.; Shahid, S.; Wang, X. Seasonal drought pattern changes due to climate variability: Case study in Afghanistan. *Water* **2019**, *11*, 1096. [CrossRef]

40. Kim, J.E.; Yoo, J.; Chung, G.H.; Kim, T.W. Hydrologic Risk Assessment of Future Extreme Drought in South Korea Using Bivariate Frequency Analysis. *Water* **2019**, *11*, 2052. [CrossRef]

41. Kwon, M.; Sung, J.H. Changes in future drought with HadGEM2-AO projections. *Water* **2019**, *11*, 312. [CrossRef]

42. Gao, Q.G.; Sombutmounvong, V.; Xiong, L.; Lee, J.H.; Kim, J.S. Analysis of Drought-Sensitive Areas and Evolution Patterns through Statistical Simulations of the Indian Ocean Dipole Mode. *Water* **2019**, *11*, 1302. [CrossRef]

43. Gao, Q.; Kim, J.S.; Chen, J.; Chen, H.; Lee, J.H. Atmospheric Teleconnection-Based Extreme Drought Prediction in the Core Drought Region in China. *Water* **2019**, *11*, 232. [CrossRef]

44. Yeh, H.F.; Hsu, H.L. Stochastic Model for Drought Forecasting in the Southern Taiwan Basin. *Water* **2019**, *11*, 2041. [CrossRef]
45. Zhang, M.; Wang, J.; Zhou, R. Attribution Analysis of Hydrological Drought Risk Under Climate Change and Human Activities: A Case Study on Kuye River Basin in China. *Water* **2019**, *11*, 1958. [CrossRef]

Article

Future Hydrological Drought Risk Assessment Based on Nonstationary Joint Drought Management Index

Jisoo Yu [1], Tae-Woong Kim [2] and Dong-Hyeok Park [1,*]

[1] Department of Civil and Environmental Engineering, Hanyang University, Seoul 04763, Korea; jisoo91@hanyang.ac.kr

[2] Department of Civil and Environmental Engineering, Hanyang University, Ansan 15588, Korea; twkim72@hanyang.ac.kr

* Correspondence: smilehyuki@hanyang.ac.kr

Received: 27 January 2019; Accepted: 8 March 2019; Published: 14 March 2019

Abstract: As the environment changes, the stationarity assumption in hydrological analysis has become questionable. If nonstationarity of an observed time series is not fully considered when handling climate change scenarios, the outcomes of statistical analyses would be invalid in practice. This study established bivariate time-varying copula models for risk analysis based on the generalized additive models in location, scale, and shape (GAMLSS) theory to develop the nonstationary joint drought management index (JDMI). Two kinds of daily streamflow data from the Soyang River basin were used; one is that observed during 1976–2005, and the other is that simulated for the period 2011–2099 from 26 climate change scenarios. The JDMI quantified the multi-index of reliability and vulnerability of hydrological drought, both of which cause damage to the hydrosystem. Hydrological drought was defined as the low-flow events that occur when streamflow is equal to or less than Q80 calculated from observed data, allowing future drought risk to be assessed and compared with the past. Then, reliability and vulnerability were estimated based on the duration and magnitude of the events, respectively. As a result, the JDMI provided the expected duration and magnitude quantities of drought or water deficit.

Keywords: climate change; drought; GAMLSS; nonstationarity

---

1. Introduction

The stationarity assumption in traditional hydrological analysis indicates that statistical characteristics of data do not vary over time. However, handling hydrological data becomes more challenging because the effects of global climate change have led to the nonstationarity of hydrometeorological variables. If the nonstationarity of hydrometeorological variables is not fully considered, the statistical results would be invalid in practice [1], especially when using climate change scenarios.

Therefore, many researchers have been focusing on identifying the time-varying characteristics. Nonstationarity can be simply checked by the trend test and change point analysis [2–4], however, the most common way of accounting for nonstationarity is the time-varying moment method [5]. This method assumes that the distribution parameters are functions of time, although variables of interest remain the same [6–9]. Rigby and Stasinopoulos [10] developed generalized additive models in location, scale, and shape (GAMLSS). GAMLSS overcomes the limitations of generalized linear models (GLM) and generalized additive models (GAM), by expanding the types of distributions and their parameters. Furthermore, it allows modeling of not only the mean and location but also skewness and kurtosis, and introduces the parametric, semi-parametric, non-parametric, or random effect terms of explanatory variables.

The GAMLSS framework has been applied to hydrological frequency analysis to provide the design criteria for managing future drought risk [11,12]. Bivariate frequency analysis based on the GAMLSS–copula model was also conducted and proved useful in hydrological prediction [13–15]. GAMLSS can be useful to estimate nonstationary index. For example, Wang et al. [16] proposed the time-dependent standardized precipitation index and Bazrafshan and Hejabi [17] suggested using the nonstationary reconnaissance drought index.

This study performed statistical analyses with 26 climate change scenarios for the Soyang River basin in South Korea to construct the bivariate time-varying copula models based on GAMLSS theory. Copula functions were used to combine the drought information estimated from low-flow events, which were reliability and vulnerability. Then a new drought index, joint drought management index (JDMI), was developed based on the nonstationary copula model. The potential drought or water supply failure events were quantified as durations and magnitudes by JDMI-based risk assessment.

2. Study Area and Data

The climate of South Korea has large seasonal variability in temperature and precipitation, with hot and humid summers and cold and dry winters. The range of those characteristics has been changing under the global climate change effects. The Intergovernmental Panel on Climate Change (IPCC) determined the emission density of greenhouse gases in the form of representative concentration pathway (RCP) scenarios [18]. In particular, under RCP 8.5, temperature and precipitation are projected to increase by 4.73 °C and 19.1%, respectively, more than the current average values in the late 21st century [19,20].

The target area is the Soyang River basin in South Korea described in Figure 1. It is one of the mid-sized basins located in the upper stream of Han River. Past observed daily streamflow data of the Soyang River basin were collected during 1976–2005, and future daily streamflow data were simulated according to precipitation estimates from the climate change scenarios for the period of 2011–2099 using the Precipitation Runoff Modeling System (PRMS). The PRMS was developed by United States Geological Survey for use in runoff analysis [21]. The 26 climate change scenarios using different Global Climate Models (GCMs) following RCP 8.5 were applied in this study and their details are depicted in Table 1. The statistical downscaling of GCMs was performed by Eum and Cannon [22].

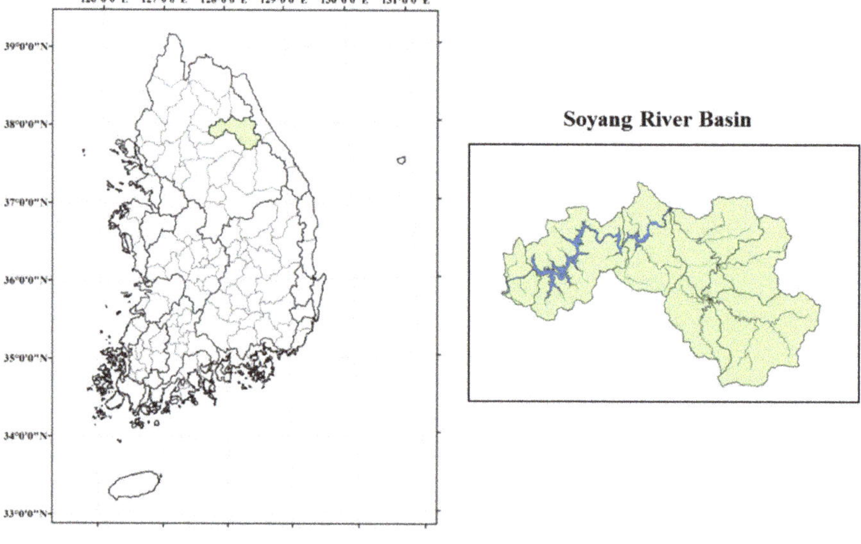

Figure 1. Study area.

Table 1. Information of climate change scenarios.

No.	GCMs	Resolution	Institution
1	BCC-CSM1-1	2.813 × 2.791	Beijing Climate Center, China Meteorological Administration
2	BCC-CSM1-1-M	1.125 × 1.122	
3	CanESM2	2.813 × 2.791	Canadian Centre for Climate Modelling and Analysis
4	CCSM4	1.250 × 0.942	Centro Euro-Mediterraneo per I Cambiamenti Climatici
5	CESM1-BGC	1.250 × 0.942	National Center for Atmospheric Research
6	CESM1-CAM5	1.250 × 0.942	
7	CMCC-CM	0.750 × 0.748	Centro Euro-Mediterraneo per I Cambiamenti Climatici
8	CMCC-CMS	1.875 × 1.865	
9	CNRM-CM5	1.406 × 1.401	Centre National de Recherches Meteorologiques
10	FGOALS-s2	2.813 × 1.659	LASG, Institute of Atmospheric Physics, Chinese Academy of Sciences
11	GFDL-ESM2G	2.500 × 2.023	Geophysical Fluid Dynamics Laboratory
12	GFDL-ESM2M	2.500 × 2.023	
13	HadGEM2-AO	1.875 × 1.250	Met Office Hadley Centre
14	HadGEM2-CC	1.875 × 1.250	
15	HadGEM2-ES	1.875 × 1.250	
16	INM-CM4	2.000 × 1.500	Institute for Numerical Mathematics
17	IPSL-CM5A-LR	3.750 × 1.895	Institut Pierre-Simon Laplace
18	IPSL-CM5A-MR	2.500 × 1.268	
19	IPSL-CM5B-LR	3.750 × 1.895	
20	MIROC5	1.406 × 1.401	Atmosphere and Ocean Research Institute (The University of Tokyo)
21	MIROC-ESM	2.813 × 2.791	Japan Agency for Marine-Earth Science and Technology, Atmosphere and Ocean Research Institute (The University of Tokyo), and
22	MIROC-ESM-CHEM	2.813 × 2.791	National Institute for Environmental Studies
23	MPI-ESM-LR	1.875 × 1.865	Max Planck Institute for Meteorology (MPI-M)
24	MPI-ESM-MR	1.875 × 1.865	
25	MRI-CGCM3	1.125 × 1.122	Meteorological Research Institute
26	NorESM1-M	2.500 × 1.895	Norwegian Climate Centre

3. Methodology

3.1. Definition of Low-Flow Event and Water Supply Performance Indices

A flow duration curve (FDC) that shows the percent of times that a given discharge is exceeded, or equaled, is widely applied for the threshold level of the streamflow drought definition [23–26]. In this study, the 80th percentile of FDC was chosen as the threshold using past streamflow data. Low streamflow provoked by the deficit of precipitation may cause a water supply failure.

The future period (2011–2099) was divided into 85 blocks using a 5-year moving window (MV), i.e., MV1 (2011–2015), MV2 (2012–2016), and so on. The characteristics of low-flow events were defined in each moving window as reliability (*rel*) and vulnerability (*vul*), as proposed by Hashimoto et al. [27] to describe the performance of water resource systems under drought conditions.

Reliability is originally defined as the probability that the desired water demand is fully supplied to users within the design period. However, in this study, reliability was expressed as Equation (1), which is the ratio of the failure period and the total design period. Vulnerability refers to the quantitative risk of water supply failure, that is, the potential magnitude or damage of failure events. Vulnerability can be estimated as the mean of the total deficit during failures as shown in Equation (2).

$$rel = \sum_{n=1}^{N} d_n / T, \tag{1}$$

$$vul = \sum_{n=1}^{N} s_n / N, \tag{2}$$

where T is the total design period and N is the number of event occurrences during design period. In this study, T is 1825 days because a 5-year moving window was used for the design period. d_n is the duration and s_n is the total deficit or magnitude of the nth low-flow event.

3.2. Time-Varying Copula Model Using GAMLSS

The nonstationary analysis may be modeled assuming the parameters are time-dependent. The mean and standard deviation are generally employed for time-varying modeling, however, shape parameters are usually considered to be constant coefficients. Rigby and Stasinopoulos [10] proposed GAMLSS that can involve two to four parameters, i.e., location (mean), scale (standard deviation), and shape (skewness, kurtosis) parameters.

A GAMLSS assumes independent observations y_i for $i = 1, 2$ ($y_1 = \{rel_1, rel_2, \cdots, rel_p\}$ and $y_2 = \{vul_1, vul_2, \cdots, vul_p\}$) with probability function $f(y_i|\Theta_i)$. $\Theta_i = (\mu_i, \sigma_i)$ is a vector of distribution parameters, which are $\mu_i = \{\mu_1, \mu_2, \cdots, \mu_p\}$ and $\sigma_i = \{\sigma_1, \sigma_2, \cdots, \sigma_p\}$. Each parameter can be related to explanatory variables. Since we assumed the drought risk to be time-dependent and induced by climate change, time was selected as an explanatory variable in the study. Explanatory variable t_k, $g(\cdot)$ is a known link function relating Θ to t_k given by Equation (3) without an additive term of random effects.

$$g_k(\theta_k) = \eta_k = \beta t_k,\tag{3}$$

where β is the vector of estimated coefficient of time-varying parameters and the length of β can be two or four, described as Equation (5). The explanatory variable time $t_k = \{t_1, t_2, \cdots, t_p\} = \{1, 2, \cdots, p\}$ is the number of the moving window, thus p is the length of the datasets (reliability and vulnerability).

The joint cumulative distribution function (JCDF) is expressed using copula of two random variables of reliability and vulnerability, as shown in Equation (4). The implementation of the time-varying copula model includes three major steps, as shown in Figure 2: (1) The estimation of reliability and vulnerability, (2) the estimation of the marginal distribution, and (3) the modeling of the copula parameter depending on the margins.

The parameters of marginal distributions (Θ_i) and copula (θ) were assumed to be constant, linear, or polynomial function as Equation (5). The parametric vector β_k were estimated within the GAMLSS framework by maximizing the log–likelihood function for copula with continuous margins written as Equation (6) [28].

$$F(y_1, y_2|\vartheta) = C(F_1(y_1|\mu_1, \sigma_1), F_2(y_2|\mu_2, \sigma_2); \theta),\tag{4}$$

$$\eta_k = const.,\tag{5a}$$

$$\eta_k = \beta_0 + \beta_1 t,\tag{5b}$$

$$\eta_k = \beta_0 + \beta_1 t + \beta_2 t^2 + \beta_3 t^3,\tag{5c}$$

$$\log L(\vartheta) = \sum_{k=1}^{p} \log\{c(F_1(y_{1k}|\mu_{1k}, \sigma_{1k}), F_2(y_{2k}|\mu_{2k}, \sigma_{2k}); \theta_k\} + \sum_{k=1}^{p}\sum_{i=1}^{2} \log\{f_i(y_{ik}|\mu_{ik}, \sigma_{ik})\}\tag{6}$$

where C and c are joint cumulative distribution function and joint probability density function estimated by copula, respectively. $\vartheta = \{\Theta_1, \Theta_2, \theta\}$ is a set of distribution parameters used in a time-dependent copula model.

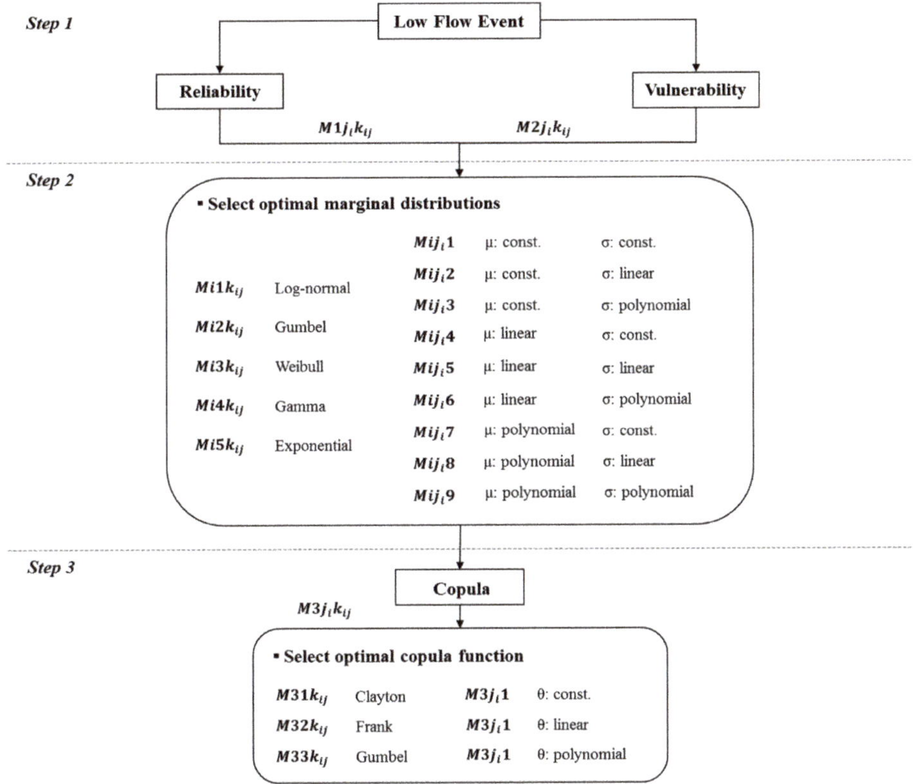

Figure 2. The time-dependent copula modeling procedure and model description.

3.3. Joint Drought Management Index (JDMI)

From Equation (4), the JCDF with bivariate marginal, $u = F_1(y_1|\Theta_1)$ and $v = F_2(y_2|\Theta_2)$ is expressed as shown in Equation (7). The cumulative probability q is usually treated as an indicator of a destructive event, thus a smaller q implies a less-dry condition. Therefore, it would be of interest in the risk analysis to know what the probability is for a random event. In this context, Salvadori and De Michele [29] adopted Kendall distribution of Archimedean copulas.

A Kendall distribution is useful to project multivariate information onto a single dimension as described as Equation (8). The copula function can have the same quantiles with different combinations of variables, e.g., $C(u,v) = q$ and $C(u^*,v^*) = q$. Therefore, it is difficult to specify the drought characteristics with the copula quantile. Since the JDMI implied the joint probability of nonexceedance of water supply performance indices, it is important to capture the ranges of sets of variables that can produce the same quantile.

In this study, the JDMI was designed to increase when reliability and vulnerability increase and vice versa. In other words, the larger the JDMI, the higher the risk. From the probability of Kendall distribution shown in Equation (8), the JDMI can be defined as shown in Equation (9).

$$C(u,v) = P[Y_1 \leq rel^*, Y_2 \leq vul^*] = C[F_1(Y_1|\Theta_1), F(Y_2|\Theta_2)] = q, \tag{7}$$

$$K_c(q) = P[C(u,v) \leq q], \tag{8}$$

$$JDMI = \phi^{-1}(K_c(q)), \tag{9}$$

where *rel** and *vul** are random values of reliability and vulnerability, respectively. ϕ is the standard normal distribution function.

4. Results and Discussion

4.1. Drought Events and Water Supply Performance Indices

The drought was defined as a period in which flow is equal or less than Q80 calculated from observed data, so that future drought risk could be assessed and compared with the past. Water supply performance indices were estimated based on low-flow drought events. The Mann–Kendall (MK) test was employed to find the monotonic trend in reliability and vulnerability estimated in 85 moving windows.

The monotonic trend was observed in 15 scenarios for reliability and vulnerability, but the scenarios were not identical. The Z-statistics of the MK test suggested that there was a significant decreasing trend in reliability in the 20 scenarios with the significance level of 0.05. A decreasing trend for vulnerability that was observed in 16 scenarios, while a significant increasing trend was observed in 10 scenarios. Thus, it is expected that the duration and the magnitude of future hydrological droughts would be less than those of the past.

The multi-index using water supply performance indices have already been tested in previous research. The drought management index (DMI) was developed using resilience and vulnerability, that account for the ability of the system to recover from drought damage [30,31]. On the contrary, the JDMI inspired by the DMI is estimated based on D and S. Thus, the JDMI can reflect the drought aspects that cause damage to the system.

4.2. Model Selection and Parameter Estimation

Five probability distributions in Table 2, i.e., log-normal, Gumbel, Weibull, Gamma, and exponential distribution, were explored to define the appropriate marginal distributions of reliability and vulnerability. In addition to investigating the relationship between the two margins, three copula functions in Table 3, i.e., Clayton, Frank, and Gumbel copula, were tested. The parameters of marginal distributions and copula were assumed to be a function of the explanatory variable, which is time, t.

Table 2. Definition and properties of the marginal distributions used in the study.

Distribution	PDF $f_m(y_m \mid \mu_m, \sigma_m)$	Link Function $g_m(\mu_m)$	$g_m(\sigma_m)$
Log-normal	$\dfrac{1}{y\sigma\sqrt{2\pi}} \exp\left[-\dfrac{\{\log(y)-\mu\}^2}{2\sigma^2}\right]$	$\log(\mu_m)$	$\log(\sigma_m)$
Gumbel	$\dfrac{1}{\sigma}\exp\left\{\left(\dfrac{y-\mu}{\sigma}\right) - \exp\left(\dfrac{y-\mu}{\sigma}\right)\right\}$	μ_m	$\log(\sigma_m)$
Weibull	$\dfrac{\sigma}{\mu}\left(\dfrac{y}{\mu}\right)^{\sigma-1}\exp\left\{-\left(\dfrac{y}{\mu}\right)^{\sigma}\right\}$	$\log(\mu_m)$	$\log(\sigma_m)$
Gamma	$\dfrac{1}{(\mu\sigma^2)^{1/\sigma^2}}\dfrac{y^{1/\sigma^2-1}\exp(-y/\mu\sigma^2)}{\Gamma(1/\sigma^2)}$	$\log(\mu_m)$	$\log(\sigma_m)$
Exponential	$\dfrac{1}{\mu}e^{-\frac{y}{\mu}}$	$\log(\mu_m)$	—

Table 3. Definition and properties of the Archimedean copula used in the study ($\varepsilon = 10^{-7}$).

Copula	CDF	Link Function
	$C(u,v;\theta)$	$g(\theta)$
Clayton	$\left(u^{-\theta} + v^{-\theta} - 1\right)^{-\frac{1}{\theta}}$	$\log(\theta - \varepsilon)$
Frank	$-\dfrac{1}{\theta}\log\left\{\dfrac{1 + (e^{-\theta u} - 1)(e^{-\theta v} - 1)}{e^{-\theta} - 1}\right\}$	θ
Gumbel	$\exp\left[-\left\{(-\log u)^{\theta} + (-\log v)^{\theta}\right\}^{\frac{1}{\theta}}\right]$	$\log(\theta - 1 - \varepsilon)$

The time-dependent copula model was fitted by the nonstationary marginal distributions. The parameters of the margins and copula were assumed to be constant, linear, and polynomial functions with the design parameters and time as described in Equations 5a, 5b, and 5c, respectively. If the parameters are independent from time, the model will be stationary with constant parameters.

Table 4 represents the result of model selection. It was decided that the distributions of best fit for reliability and vulnerability were the Weibull and log-normal, respectively. The p-value of the Kolmogorov–Smirnov (KS) test for each marginal distribution supported the model decision with 5% significance level. Finally, the most appropriate copula models were determined based on the Akaike information criterion (AIC), and Gumbel was dominantly selected.

Table 4. Time-dependent copula model selection of each scenarios.

SCN	Reliability		P-KS	Vulnerability		P-KS	Copula		AIC
	Distribution			Distribution			Distribution		
S#01	Weibull	M134	0.31	Weibull	M231	0.91	Clayton	M313	455.21
S#02	Weibull	M132	0.48	log-normal	M215	0.09	Gumbel	M333	430.79
S#03	Weibull	M131	0.93	Weibull	M235	0.62	Frank	M323	428.96
S#04	Gamma	M141	0.76	log-normal	M211	1.00	Gumbel	M333	443.54
S#05	Weibull	M137	0.67	log-normal	M213	0.20	Gumbel	M333	444.26
S#06	log-normal	M113	0.97	log-normal	M211	0.73	Clayton	M313	414.55
S#07	Weibull	M134	0.51	log-normal	M214	0.33	Gumbel	M333	413.12
S#08	Gamma	M144	0.46	log-normal	M214	0.18	Clayton	M313	472.41
S#09	Weibull	M131	0.86	log-normal	M211	0.98	Gumbel	M333	415.80
S#10	Gamma	M147	0.33	log-normal	M212	0.68	Clayton	M313	443.71
S#11	Weibull	M131	0.67	Gamma	M241	0.80	Frank	M323	446.45
S#12	Weibull	M135	0.11	log-normal	M214	0.34	Gumbel	M333	419.45
S#13	Weibull	M131	0.54	log-normal	M212	0.60	Gumbel	M333	439.64
S#14	Gamma	M142	0.10	Gamma	M247	0.22	Clayton	M313	405.54
S#15	Weibull	M138	0.75	Gamma	M241	0.26	Clayton	M313	417.80
S#16	Gamma	M145	0.28	log-normal	M213	0.35	Gumbel	M333	419.98
S#17	Weibull	M135	0.56	Gamma	M247	0.75	Clayton	M313	429.78
S#18	Weibull	M137	0.92	Gamma	M241	1.00	Frank	M323	420.80
S#19	Weibull	M134	0.63	log-normal	M211	0.34	Frank	M323	447.95
S#20	log-normal	M111	0.76	Gamma	M247	0.07	Clayton	M313	399.52
S#21	log-normal	M113	0.54	Gumbel	M224	0.42	Clayton	M313	443.96
S#22	Weibull	M133	0.88	Weibull	M232	0.95	Frank	M323	412.28
S#23	log-normal	M111	0.90	Gamma	M241	0.39	Frank	M323	419.94
S#24	Gamma	M144	0.60	log-normal	M211	0.49	Gumbel	M333	415.35
S#25	Gamma	M145	0.08	log-normal	M214	0.64	Gumbel	M333	397.11
S#26	Gamma	M145	0.27	Gamma	M243	0.23	Gumbel	M333	446.31

* P-KS is the p-value of the Kolmogorov–Smirnov (KS) test; if the p-value is larger than a significance level (0.05), the result of distribution selection is valid.

The stationary copula models were also estimated for comparison. Figure 3 describes the AIC comparison between the stationary and nonstationary copula models. The AIC values of nonstationary models were estimated to be lower than those of stationary models in all scenarios. The lower AIC indicates a better model, thus, the time-dependent copula would be more appropriate in the study area. It should be noted that the marginal distributions and the copula of the stationary and nonstationary model were not identical; log-normal was selected for all of the margins and Gumbel copula was dominant for the stationary model.

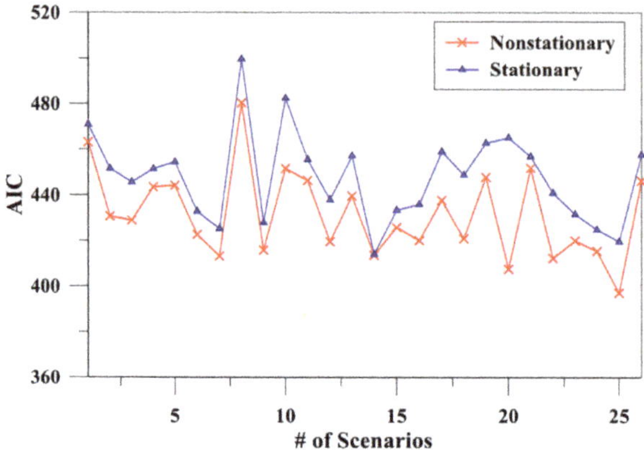

Figure 3. Akaike information criterion (AIC) comparison of nonstationary and stationary copula model.

4.3. Risk Assessment Based on Nonstationary JDMI

The drought event was defined when both reliability and vulnerability were larger than average values in a probabilistic approach. The droughts based on JDMI can be classified into five states, i.e., normal (D0), mild (D1), moderate (D2), severe (D3), and extreme (D4). Figure 4 shows the drought state classification based on the JDMI estimated for all scenarios.

The future period was divided into three time domains to clarify drought occurrence patterns and their characteristics. The periods of P1 (2011–2040), P2 (2041–2070), and P3 (2071–2099) can be substituted to the moving windows, which are MV1–30, MV31–60, and MV61–85, respectively. In other words, each period consists of about 30 different 5-year moving windows: (1) P1: MV1 (2011–2015) to MV30 (2040–2044); (2) P2: MV31 (2041–2045) to MV60 (2070–2074); (3) P3: MV61 (2071–2075) to MV85 (2095–2099).

As shown in Figure 4, the extreme drought (D4) rarely occurred but only five events were observed in all scenarios (scenario #03, #6, #08, #21, and #23), and appeared in P2 and P3. The number of droughts was larger in P1 compared with P2 and P3, however, the mild and moderate droughts (D1 and D2) were concentrated in P1. Among the scenarios, results from the severe and extreme drought (D3 and D4) occurrences were slightly more frequent in P2 and P3 than P1. Nevertheless, it was obvious that drought appearance declined in P2 and P3; thus, P1 would be more prone to water deficit.

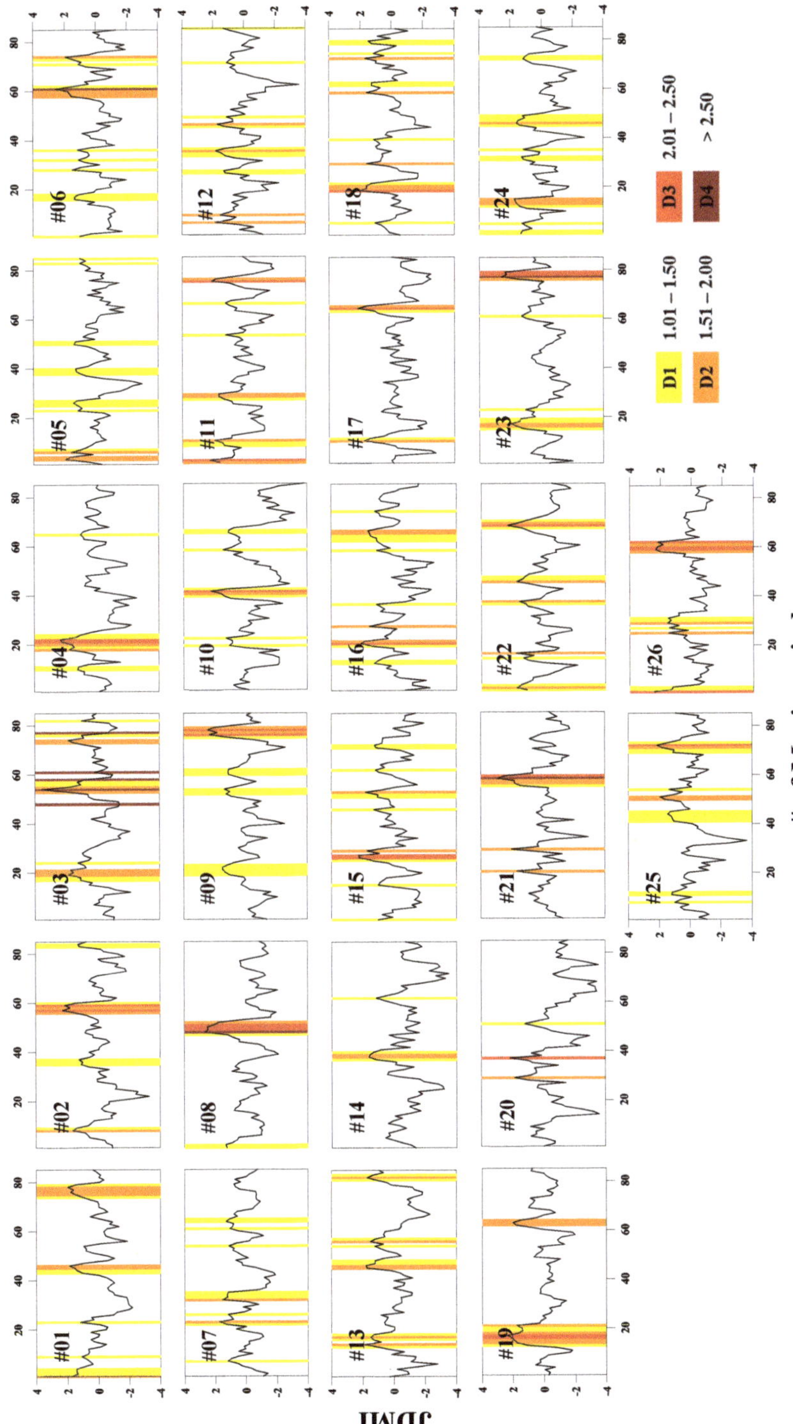

Figure 4. Joint drought management index (JDMI) estimation and drought state classification of 26 scenarios.

Table 5 depicts the maximum drought events in all scenarios and the corresponding values of reliability and vulnerability. Reliability and vulnerability could be representative of drought duration and magnitude and, thus, we can deduce the drought characteristics from those indexes. In the S#01, for example, reliability and vulnerability were estimated as 0.46 and 570.15 m^3/s based on JDMI of 1.98. During a 5-year moving window, the expected drought occurrence duration would be 0.46 × 365 (days) × 5 (years) = 839.5 days. Also, the expected maximum drought magnitude would be 570.15 × 24 (h) × 60 (min) × 60 (s) = 49,260,960 m^3 per day, due to the daily streamflow data basis.

Table 5. Maximum events based on JDMI of each scenario and the corresponding values of reliability and vulnerability.

SCN	# of Moving Window (year)		JDMI	Reliability	Vulnerability (m^3/s)
S#01	78	(2088–2092)	1.98	0.46	570.15
S#02	57	(2067–2071)	2.28	0.39	929.20
S#03	54	(2064–2068)	3.56	0.32	780.31
S#04	22	(2032–2036)	2.36	0.40	1078.86
S#05	3	(2013–2017)	1.98	0.33	550.76
S#06	61	(2071–2075)	2.56	0.31	926.85
S#07	23	(2033–2037)	1.73	0.34	563.84
S#08	48	(2058–2062)	2.64	0.43	1273.97
S#09	78	(2088–2092)	2.40	0.42	783.54
S#10	41	(2051–2055)	2.18	0.50	892.56
S#11	2	(2012–2016)	2.19	0.39	838.17
S#12	6	(2016–2020)	1.86	0.25	781.09
S#13	14	(2024–2028)	1.93	0.34	749.63
S#14	38	(2048–2052)	1.62	0.35	792.17
S#15	27	(2037–2041)	2.26	0.33	685.77
S#16	20	(2030–2034)	2.14	0.31	778.84
S#17	64	(2074–2078)	2.15	0.40	816.04
S#18	19	(2029–2033)	2.03	0.36	743.18
S#19	17	(2027–2031)	2.39	0.35	814.58
S#20	37	(2047–2051)	2.09	0.37	992.74
S#21	58	(2068–2072)	2.89	0.37	1075.18
S#22	68	(2078–2082)	2.25	0.33	908.85
S#23	77	(2087–2091)	2.53	0.36	704.78
S#24	15	(2025–2029)	1.73	0.21	545.02
S#25	72	(2082–2086)	2.12	0.29	497.73
S#26	1	(2011–2015)	2.39	0.36	633.40

The JDMI can quantify the potential hydrological drought characteristics that would cause severe water shortage every five years. According to Table 5, the corresponding values of reliability and vulnerability of each maximum JDMI ranged from 0.21 to 0.50, and from 497.73 to 1273.97 m^3/s, respectively. Therefore, it is expected for the Soyang River basin that the maximum drought damage would last at least 386 to 915 days and induce a water deficit from 43 × 10^6 to maximum 110 × 10^6 m^3 for an event during five years (represented as 5-year moving window in the study).

5. Summaries and Conclusions

This study proposed the nonstationary JDMI built under the time-dependent copula model to assess the potential risk of hydrological droughts that could induce water deficit. The focus was on the nonstationarity in the parameters of distribution function of margins and copula detected using the GAMLSS framework. The appropriate model selection procedure was generally carried out based on the log–likelihood functions and AIC.

The data corresponding to 26 climate change scenarios of Soyang River basin of were used in the analysis. The future low-flow drought was defined using the future simulated data, but the threshold level was estimated from past data. The water supply performance indexes, reliability and vulnerability, were respectively calculated from the duration and magnitude of the drought events occurring in 5-year moving windows.

Not all marginal distributions were found to be nonstationary based on the MK test, however, some of the datasets showed best fitting when using the time-varying parameter, even when they were under a stationary condition. It should be noted that among 11 datasets of reliability and vulnerability revealed not to have a trend, the constant parameters showed the best fitting for only 5 and 6 datasets, respectively. On the other hand, the copula with the polynomial function of time for the dependence parameter had better performance than the one with the constant or linear functions in all datasets. Therefore, we can conclude that the time-varying model is more suitable for the study area.

An appropriate drought index can serve as the important base in drought management, preparation, and mitigation. The nonstationary JDMI may provide the expected quantities of drought risk under changing environmental conditions. Therefore, it is possible to establish the targeted safety level through reliability and vulnerability, which are corresponding values of JDMI in the specific period. Water supply safety assessment is a priority for drought response/mitigation and also water supply system design.

However, the nonstationary model developed in this research does not include other hydrometeorological variables, with only time as an explanatory variable. Thus, it cannot fully consider the nonstationary behaviors of drought characteristics. However, the multi-index would increase uncertainty by combining the variables in a probabilistic approach. Nevertheless, future studies should establish new drought indexes with climate variables, i.e., temperature, precipitation, and evaporation.

Author Contributions: J.Y.: Methodology, analysis, and writing; T.-W.K.: funding acquisition and review; D.-H.P.: conceptualization and data collection.

Funding: This work is supported by the Korea Environmental Industry & Technology Institute (KEITI) grant funded by the Ministry of Environment (Grant 18AWMP-B083066-05).

Conflicts of Interest: The authors declare no conflict of interest.

References

1. Milly, P.C.D.; Betancourt, J.; Falkenmark, M.; Hirsch, R.M.; Kundzewicz, Z.W.; Lettenmaier, D.P.; Stouffer, R.J. Stationarity is Dead: Whither Water Management? *Science* **2008**, *319*, 573–574. [CrossRef]
2. Kendall, M.G. Rank correlation methods. *Biometrika* **1957**, *44*, 298. [CrossRef]
3. Mann, H.B. Nonparametric Tests Against Trend. *Econometrica* **1945**, *13*, 245–259. [CrossRef]
4. Pettitt, A.N. A Non-Parametric Approach to the Change-Point Problem. *Appl. Stat.* **1979**, *28*, 126–135. [CrossRef]
5. Strupczewski, W.G.; Singh, V.P.; Feluch, W. Non-stationary approach to at-site flood frequency modelling I. Maximum likelihood estimation. *J. Hydrol.* **2001**, *248*, 123–142. [CrossRef]
6. El Adlouni, S.; Ouarda, T.B.M.J.; Zhang, X.; Roy, R.; Bobée, B. Generalized maximum likelihood estimators for the nonstationary generalized extreme value model. *Water Resour. Res.* **2007**, *43*. [CrossRef]
7. Read, L.K.; Vogel, R.M. Reliability, return periods, and risk under nonstationarity. *Water Resour. Res.* **2015**, *51*, 6381–6398. [CrossRef]
8. Sugahara, S.; da Rocha, R.P.; Silveira, R. Non-stationary frequency analysis of extreme daily rainfall in Sao Paulo, Brazil. *Int. J. Climatol.* **2009**, *29*, 1339–1349. [CrossRef]
9. Wi, S.; Valdés, J.B.; Steinschneider, S.; Kim, T.W. Non-stationary frequency analysis of extreme precipitation in South Korea using peaks-over-threshold and annual maxima. *Stoch. Environ. Res. Risk Assess.* **2015**, *30*, 583–606. [CrossRef]
10. Rigby, R.A.; Stasinopoulos, D.M. Generalized additive models for location, scale and shape. *J. R. Stat. Soc. Ser. C-Appl. Stat.* **2005**, *54*, 507–544. [CrossRef]
11. Du, T.; Xiong, L.; Xu, C.Y.; Gippel, C.J.; Guo, S.; Liu, P. Return period and risk analysis of nonstationary low-flow series under climate change. *J. Hydrol.* **2015**, *527*, 234–250. [CrossRef]

12. Yan, L.; Xiong, L.; Guo, S.; Xu, C.-Y.; Xia, J.; Du, T. Comparison of four nonstationary hydrologic design methods for changing environment. *J. Hydrol.* **2017**, *551*, 132–150. [CrossRef]

13. Ahn, K.H.; Palmer, R.N. Use of a nonstationary copula to predict future bivariate low flow frequency in the Connecticut river basin. *Hydrol. Process.* **2016**, *30*, 3518–3532. [CrossRef]

14. Jiang, C.; Xiong, L.; Wang, D.; Liu, P.; Guo, S.; Xu, C.Y. Separating the impacts of climate change and human activities on runoff using the Budyko-type equations with time-varying parameters. *J. Hydrol.* **2015**, *522*, 326–338. [CrossRef]

15. Zhang, T.; Wang, Y.; Wang, B.; Tan, S.; Feng, P. Nonstationary Flood Frequency Analysis Using Univariate and Bivariate Time-Varying Models Based on GAMLSS. *Water* **2018**, *10*, 819. [CrossRef]

16. Wang, Y.; Li, J.; Feng, P.; Hu, R. A Time-Dependent Drought Index for Non-Stationary Precipitation Series. *Water Resour. Manag.* **2015**, *29*, 5631–5647. [CrossRef]

17. Bazrafshan, J.; Hejabi, S. A Non-Stationary Reconnaissance Drought Index (NRDI) for Drought Monitoring in a Changing Climate. *Water Resour. Manag.* **2018**, *32*, 2611–2624. [CrossRef]

18. Intergovernmental Panel on Climate Change. *Climate Change 2014 Mitigation of Climate Change*; Cambridge University Press: Cambridge, UK, 2014. [CrossRef]

19. Suh, M.S.; Oh, S.G.; Lee, Y.S.; Ahn, J.B.; Cha, D.H.; Lee, D.K.; Hong, S.Y.; Min, S.K.; Park, S.C.; Kang, H.S. Projections of high resolution climate changes for South Korea using multiple-regional climate models based on four RCP scenarios. Part 1: Surface air temperature. *Asia-Pac. J. Atmos. Sci.* **2016**, *52*, 151–169. [CrossRef]

20. Oh, S.G.; Suh, M.S.; Lee, Y.S.; Ahn, J.B.; Cha, D.H.; Lee, D.K.; Hong, S.Y.; Min, S.K.; Park, S.C.; Kang, H.S. Projections of high resolution climate changes for South Korea using multiple-regional climate models based on four RCP scenarios. Part 2: Precipitation. *Asia-Pac. J. Atmos. Sci.* **2016**, *52*, 171–189. [CrossRef]

21. Markstrom, S.L.; Regan, R.S.; Hay, L.E.; Viger, R.J.; Webb, R.M.; Payn, R.A.; LaFontaine, J.H. PRMS-IV, the precipitation-runoff modeling system, version 4. *US Geol. Surv. Tech. Methods* **2015**, 6-B7. [CrossRef]

22. Eum, H.I.; Cannon, A.J. Intercomparison of projected changes in climate extremes for South Korea: Application of trend preserving statistical downscaling methods to the CMIP5 ensemble. *Int. J. Climatol.* **2017**, *37*, 3381–3397. [CrossRef]

23. Beyene, B.S.; Van Loon, A.F.; Van Lanen, H.A.J.; Torfs, P.J.J.F. Investigation of variable threshold level approaches for hydrological drought identification. *Hydrol. Earth Syst. Sci. Discuss.* **2014**, *11*, 12765–12797. [CrossRef]

24. Fleig, A.K.; Tallaksen, L.M.; Hisdal, H.; Demuth, S. A global evaluation of streamflow drought characteristics. *Hydrol. Earth Syst. Sci.* **2006**, *10*, 535–552. [CrossRef]

25. Rivera, J.A.; Araneo, D.C.; Penalba, O.C. Threshold level approach for streamflow drought analysis in the Central Andes of Argentina: A climatological assessment. *Hydrol. Sci. J.* **2017**, *62*, 1949–1964. [CrossRef]

26. Tallaksen, L.M.; Madsen, H.; Clausen, B. On the definition and modelling of streamflow drought duration and deficit volume. *Hydrol. Sci. J.-J. Des. Sci. Hydrol.* **1997**, *42*, 15–33. [CrossRef]

27. Hashimoto, T.; Stedinger, J.R.; Loucks, D.P. Reliability, Resiliency, and Vulnerability Criteria for Water-Resource System Performance Evaluation. *Water Resour. Res.* **1982**, *18*, 14–20. [CrossRef]

28. Marra, G.; Radice, R. Bivariate copula additive models for location, scale and shape. *Comput. Stat. Data Anal.* **2017**, *112*, 99–113. [CrossRef]

29. Salvadori, G.; De Michele, C. Frequency analysis via copulas: Theoretical aspects and applications to hydrological events. *Water Resour. Res.* **2004**, *40*, W12511. [CrossRef]

30. Chanda, K.; Maity, R.; Sharma, A.; Mehrotra, R. Spatiotemporal variation of long-term drought propensity through reliability-resilience-vulnerability based Drought Management Index. *Water Resour. Res.* **2014**, *50*, 7662–7676. [CrossRef]

31. Maity, R.; Sharma, A.; Nagesh Kumar, D.; Chanda, K. Characterizing Drought Using the Reliability-Resilience-Vulnerability Concept. *J. Hydrol. Eng.* **2013**, *18*, 859–869. [CrossRef]

Article

Comprehensive Drought Assessment Using a Modified Composite Drought index: A Case Study in Hubei Province, China

Si Chen [1,2], Wushuang Zhong [1], Shihan Pan [3], Qijiao Xie [1,2,*] and Tae-Woong Kim [4,*]

[1] School of Resources and Environmental Science, Hubei University, Wuhan 430062, Hubei Province, China; kathryncs123@hotmail.com (S.C.); Zhongwushuang@outlook.com (W.Z.)
[2] Key Laboratory of Regional Development and Environmental Response, Wuhan 430062, Hubei Province, China
[3] Department of Civil and Environmental System Engineering, Hanyang University, Seoul 04763, Korea; panshihan1224@gmail.com
[4] Department of Civil and Environmental Engineering, Hanyang University, Ansan 15588, Korea
* Correspondence: xieqijiao@126.com (Q.X.); twkim72@hanyang.ac.kr (T.-W.K.)

Received: 30 November 2019; Accepted: 6 February 2020; Published: 9 February 2020

Abstract: Under the background of global climate change, accurate monitoring and comprehensive assessment of droughts are of great practical significance to sustain agricultural development. Considering multiple causes and the complexity of the occurrence of drought, this paper employs multiple input variables, i.e., precipitation, temperature, evaporation, and surface water content to construct a modified composite drought index (MCDI) using a series of mathematical calculation methods. The derived MCDI was calculated as a multivariate drought index to measure the drought conditions and verify its accuracy in Hubei Province in China. Compared with the existing multivariate drought index, i.e., meteorological drought composite index (CI), there was a high level of correlation in monitoring drought events in Hubei Province. Moreover, according to the drought historical record, the significant drought processes monitored by the MCDI were consistent with actual drought conditions. Furthermore, temporal and spatial analysis of drought in Hubei Province was performed based on the monitoring results of the MCDI. This paper generalizes the development of the MCDI as a new method for comprehensive assessments of regional drought.

Keywords: comprehensive drought monitoring; Hubei Province; multivariate; multisource data

---

1. Introduction

Drought is a natural hazard which has a huge impact on the world economy; its primary cause is a persistent lack of precipitation [1]. Against the background of a warming climate and increased evaporation, a large hydro-meteorological imbalance and more frequent droughts occurred, resulting in significant damages to agriculture production and human livelihoods [2]. Thus, timely and precise drought monitoring for occurrence, severity, and spatial extent plays a vital role in drought risk assessment and water resources management.

In drought monitoring, various drought indices have been developed to measure the drought characteristics, among which most were based on one of two kinds of data source. One is based on traditional meteorological observed data, such as the Standardized Precipitation Index [3], Palmer Drought Severity Index [4], and Z index [5,6]. The SPI has been extensively applied as a basic index for monitoring drought in many countries, (e.g., United States [7], China [8], and Korea [9] and European countries [10]). It is simple to calculate, and has the advantage of being able to monitor capability across time scales. The PDSI is a widely used drought index in the United States and Europe, as it considers precipitation, temperature, and soil effective water content [4]. The Z index was applied to monitor droughts in China by the National Climate Center and shows good capability for monitoring drought in Guizhou Province [11]. The ground information obtained using the aforementioned meteorological monitoring indices are not only accurate, but can also reveal the influence of environmental, anthropic, and other factors on the drought development process. However, this kind of traditional meteorological drought monitoring also has certain disadvantages; for example, in cases of sparse meteorological stations in the research area or mismatches between meteorological stations and the research area, the ground data acquisition may lead to a decline in the accuracy of results.

The other kind of drought indices are based on remote sensing data, such as surface water content index [12] and vegetation supply water index [13]. Such indices have been widely used due to their high space–time resolution and availability of monitoring drought conditions in regionally continuous locations. However, most remote-sensed data focus on a single variable, such as soil moisture or vegetation cover, which results in uncertainty and high vulnerability in monitoring development of drought in response to the combined effects of climate change [14]. In addition, compared with meteorological observed data, remote sensing monitoring is a new technology. Its short time series is the biggest challenge facing remote sensing data, along with its lack of wide applications for the large-scale monitoring of drought events [15].

Given that single variable-based drought indices do not consider a variety of drought-induced factors, they cannot fully reflect the information about drought, and may lead to inaccurate drought monitoring results [16]. Based on this, a large number of multivariate drought indices have been derived in recent years, promoting the development of comprehensive drought monitoring with advanced research methods and extensive adaptability [15]. Kao and Govindaraju proposed the Joint Drought Index (JDI) and verified it as an effective drought monitoring index [17]. Sun et al. proposed a multi-index drought (MID) model to combine the strengths of various drought indices for agricultural drought risk assessment in Canada [18]. Hao et al. designed a Meteorological Drought Index using the PCA method (PMDI) which showed good monitoring results in south-west China [19]. However, coupled with the differences in spatial and temporal scales for various drought factors, the multivariate drought indices cannot be easily applied, and it is difficult to compare the monitoring results among study areas, which makes it challenging to accurately monitor a drought.

Therefore, it is necessary to construct a more general and comprehensive drought monitoring index to compensate for the deficiencies of the current indices. This study made full use of the complementary advantages of meteorological site and remote sensing data. Among this, various drought indexes were comprehensively compared; finally, rainfall, temperature, evaporation, and surface water content index (SWCI) were chosen as the input variables to derive the modified comprehensive drought index (MCDI) considering multiple drought-related factors in drought monitoring. Hubei Province was selected as the study area in which to apply this index and to assess its accuracy in the comprehensive monitoring of regional drought.

2. Methods and Data

2.1. Study Area

Hubei Province is located in central China (29°01′N to 33°6′N, 108°21′E to 116°07′E) and has an area of 18.59×10^4 km^2, occupying 1.94% of the total area of China. The terrain features of Hubei Province are flat in the middle, surrounded by hills in the north, east, and west. Hubei Province has the typical monsoon climate of the continental eastern coast, with abundant rainfall, i.e., 800 mm to 1600 mm annually, which is greater than the average amount in the rest of China. However, due to the influence of monsoons, the temporal and spatial distributions of precipitation are irregular. In terms of time scale, precipitation in winter is 1/3 of that in summer, and the precipitation from May to September accounts for about 60% of the annual levels. Especially affected by the plum rains season, precipitation is the heaviest and lasts the longest from mid-June to mid-July. In terms of spatial scale, precipitation in the southeast is slightly higher than in the northwest. In the past 20 years, natural disasters such as floods, droughts, hurricanes, and freezing damages have occurred frequently in Hubei Province [20]. According to the aforementioned characteristics of topography, geomorphology, climate, and hydrology, the province can be divided into five areas: the humid mountainous area in the southern Hubei Province, the semi-humid plain area in the central Hubei Province, the humid hilly area in the southeastern Hubei Province, the arid mountainous area in the northwestern Hubei Province, and the arid hilly area in the northern Hubei Province.

2.2. Data

2.2.1. Selection and Preprocessing of Meteorological Data

In this paper, the observed data of 16 meteorological stations in Hubei Province from 1990 to 2017 were collected from the National Meteorological Center of China Meteorological Administration (http://data.cma.cn/), including daily rainfall, daily minimum temperature, daily maximum temperature, daily evaporation, and relative humidity. This study employed monthly cumulative precipitation, monthly average temperature, and monthly average evaporation of the 16 meteorological stations as independent meteorological variables to derive the comprehensive drought index. The spatial distribution map and location information of the 16 meteorological stations in Hubei Province are shown in Figure 1.

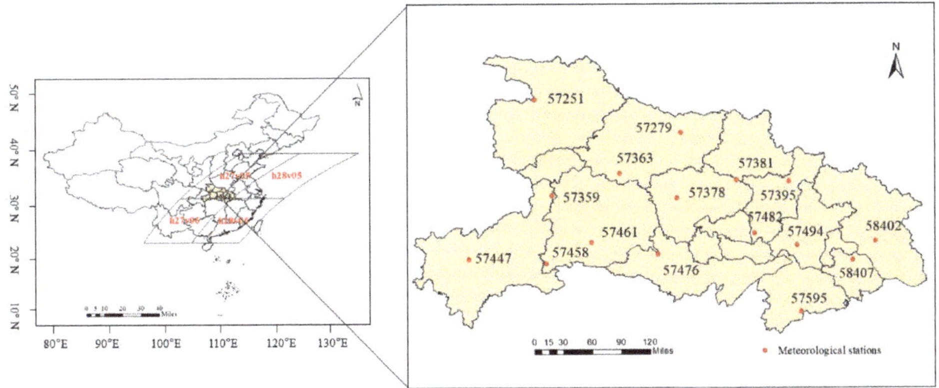

Figure 1. Distribution of Meteorological Stations in Hubei Province. (Overlay with four regions of MODIS, i.e., h27v05, h27v06, h28v05, h28v06).

2.2.2. Selection and Preprocessing of Remote Sensed Data

Moderate-resolution Imaging Spectroradiometer (MODIS) data are widely used in remote sensing research due to their high temporal resolution, wide spectral range, and moderate spatial resolution. In this study, we obtained the monthly MOD13A3, MOD11A2, and MOD09A1 data with a resolution of 250 m, 1 km, and 250 m according to the NASA portal (http://modis.gsfc.nasa.gov/data/) from 2002 to 2017. These data were used to calculate the surface water content index (SWCI) [21] to better reflect surface water content and its variations. Based on study of the spectral reflectance of water and soil, the SWCI was proposed as the following formula [13]:

$$\text{SWCI} = \frac{\rho_6 - \rho_7}{\rho_6 + \rho_7} \tag{1}$$

where ρ_6 and ρ_7 were the reflectance of bands 6 (1.628 μm–1.652 μm) and band 7 (2.105 μm–2.155 μm) from the MODIS products, respectively. As the SWCI considers the spectral reflectance of both soil and vegetation as affected by water content and can accurately monitor the shallow soil water content, this study selected it as the independent variable of remote sensing data by which to derive the comprehensive drought index. To be consistent in spatial scale with the aforementioned meteorological data, we firstly calculated the SWCI value of each pixel in Hubei Province, and then established a buffer zone centering on the spatial pixel location of the meteorological station. The average value of all pixels in the buffer zone was used as the final SWCI data for the station [22].

3. Construction of MCDI Based on MultiSource Data

Three meteorological variables (i.e., precipitation, temperature, and evaporation) and one remotely-sensed variable (SWCI) were selected to construct the modified comprehensive drought index (MCDI) to reflect drought dynamics in meteorological and agricultural systems. The drought condition measured by MCDI was then verified with historical drought records.

Construction of MCDI

Considering the similarity of input variables and good performance in comprehensive monitoring drought conditions in Han River watershed in South Korea, we followed the method of deriving the composite drought index (CDI) [23] to construct the MCDI for regional drought assessment. The input variables were monthly precipitation (P), monthly average temperature (T), evaporation (E), and SWCI from 2002 to 2017. Here, P and SWCI are negatively related to drought severity, while T and E are positively related to drought severity. The following is an overview of the step-by-step process of computing the MCDI; detailed information can be obtained from the methodology illustration in [23].

Let X be the vector of input variables $X = (x_1, x_2, x_3, x_4)$ representing monthly precipitation, SWCI, monthly average temperature, and evaporation data, respectively. For each $x_i, i = 1, 2, 3, 4$, the dataset was arranged as yearly data for each month, i.e., *for* $t = January$, $x_{it} = [x_{itj}]$, where x_{itj} indicating for the x_i value in January of the jth year, $j = (1, 2, \ldots, n)$ indicating for the first, second, $\ldots$, and the last year of the data record, and $t =$ monthly (January, February, March, $\ldots$, December). The dimensionless quantization processing procedure for a given month t was performed for x_{itj} to obtain the corresponding standardized value sx_{itj} as follows:

$$sx_{itj} = \frac{abs\left(x_{itj}\right)}{\sum_{j=1}^{n} abs\left(x_{itj}\right)} \tag{2}$$

Next, the weights corresponding to the i variables were calculated based on information entropy [24] (Equations (2)–(4)). Entropy weight was employed to balance the relationship between variables and provide unbiased relative weight according to the variability of the parameters in the database; the larger the entropy weight, the greater the change of a particular variable:

$$EM_{it} = -\frac{1}{\ln(n)} \sum_{j=1}^{n} sx_{itj} \ln\left(sx_{itj}\right) \tag{3}$$

$$DS_{it} = 1 - EM_{it} \tag{4}$$

$$Ew_{it} = \frac{DS_{it}}{\sum_{i=1}^{4} DS_{it}} \tag{5}$$

Among these variables, Ew_{it} is the weight allocated to the ith variable for month t, which needs to satisfy $\sum_{i=1}^{4} Ew_{it} = 1$.

Next, we defined the wettest condition (MWC) (Equation (5)) and the driest condition (MDC) (Equation (6)) by selecting a set of maximum values that belongs to the P and SWCI datasets and a set of minimum values that belongs to the T and E datasets.

$$MWC = \left(A_1^+, A_2^+ \dots, A_n^+\right) \tag{6}$$

$$MDC = \left(A_1^-, A_2^- \dots, A_n^-\right) \tag{7}$$

In these equations,

$$A_i^+ = \left\{max\{sx_{itj}\},\ i = 1 \text{ and } 2; min\{sx_{itj}\}, i = 3 \text{ and } 4\right\}$$

$$A_i^- = \left\{min\{sx_{itj}\}, i = 1 \text{ and } 2; max\{sx_{itj}\}, i = 3 \text{ and } 4\right\}$$

The weighted Euclidean distance [24] was used as a similarity measure to calculate the difference between the current condition and the MWC/MDC (Equation (7)):

$$D_{itj}^- = A_i^- - sx_{itj} \tag{8}$$

$$D_{itj}^+ = A_i^+ - sx_{itj} \tag{9}$$

$$D_{tj}^+ = \sqrt{\sum_{i=1}^{4} Ew_i\left(D_{itj}^+\right)^2} \tag{10}$$

$$D_{tj}^- = \sqrt{\sum_{i=1}^{4} Ew_i\left(D_{itj}^-\right)^2} \tag{11}$$

Finally, the MCDI for a given month t and year j is calculated as follows:

$$MCDI_{tj} = \frac{D_{tj}^-}{D_{tj}^- + D_{tj}^+} \tag{12}$$

where MCDI ranges from 0.0 to 1.0, and the larger the value (closer to 1.0), the wetter the condition, and vice versa. In addition, according to the criteria defined by Waseem et al. [23], the drought state was classified according to the calculated MCDI values as shown in Table 1.

Table 1. Drought Intensity Classification based on CDI value.

Drought Category	Drought State	MCDI
D_0	Non-drought	$0.4 \leq MCDI$
D_1	Slight drought	$0.3 \leq MCDI < 0.4$
D_2	Moderate drought	$0.2 \leq MCDI < 0.3$
D_3	Severe drought	$0.1 \leq MCDI < 0.2$
D_4	Extreme drought	$MCDI < 0.1$

4. Results

The monthly MCDI values of the 16 meteorological stations in Hubei Province from 2002–2017 were calculated and evaluated for performance in assessing drought.

4.1. MCDI Response to Input Variables

To investigate the response of MCDI to individual input variables, the mean monthly series of MCDI values for all stations was plotted in Figure 2, together with the plots for percentage of precipitation anomalies (PA), temperature anomalies (TA), evaporation anomalies (ETA), and SWCI monthly series.

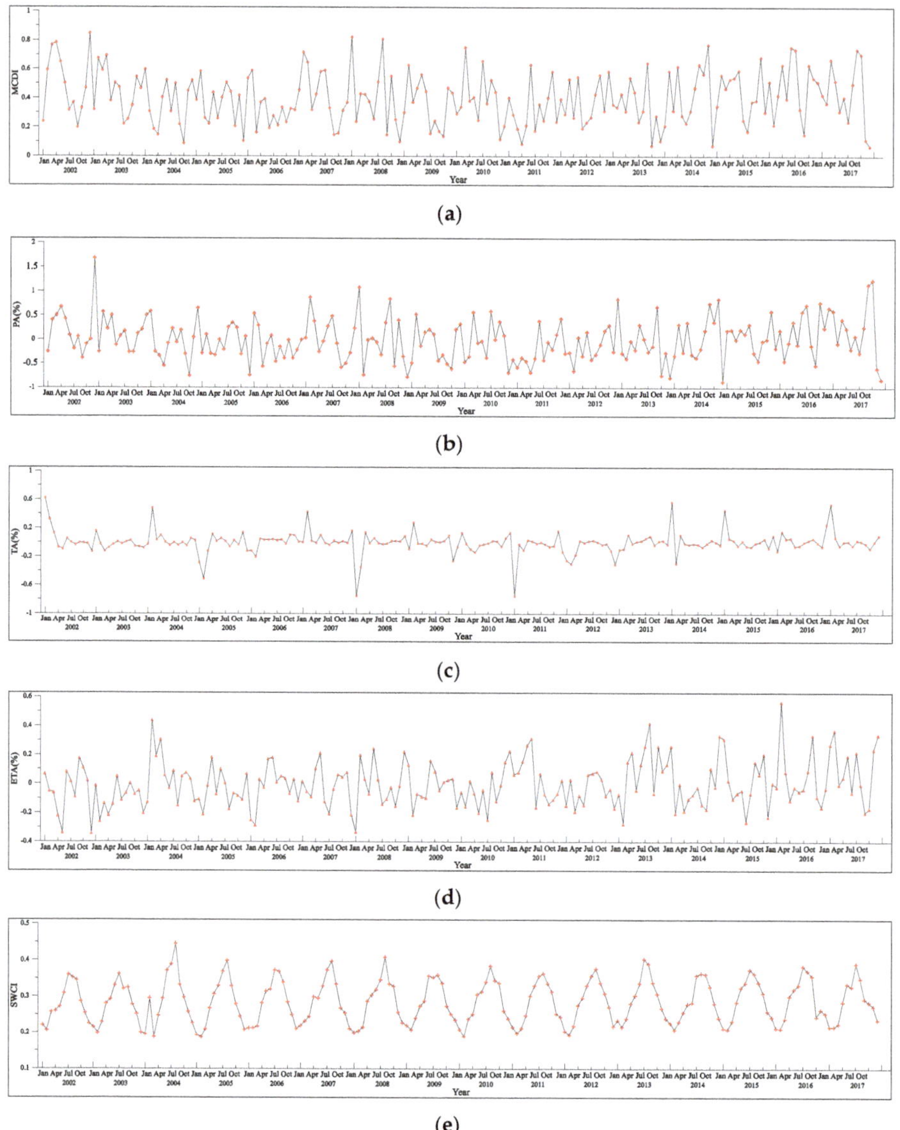

Figure 2. Monthly mean series of (**a**) MCDI, (**b**) PA, (**c**) TA, (**d**) ETA, and (**e**) SWCI for all stations in Hubei Province.

The figure showed that the MCDI series had a similar tendency to the PA, but a negative correlation with TA and ETA series. The driest months with the lowest MCDI values had very small negative PA and large positive TA and ETA values for December 2004, December 2008, and October 2013. The SWCI values seemed to weakly correlate with MCDI since they exhibited obvious seasonality, with the highest values occurring in the summer and the lowest in the winter. However, the SWCI that reflects surface water content can have a lagging effect on MCDI-based comprehensive drought assessment when meteorological conditions change; for example, in May 2010. The PA in that month indicated reduced precipitation compared to the average amount, with a value of −3%, while the SWCI showed a wet condition with a large value of 0.31. Finally, the condition in that month was classified as non-drought, based on an MCDI value of 0.43. Thus, drought assessment based on the MCDI was verified to be a result of considering the combined effects of individual input variables.

4.2. Performance Verification of MCDI-Based Drought Assessment

The meteorological drought composite index (CI) was employed as a comprehensive drought assessment index to be compared with the MCDI, as it can reflect not only short- and long-term meteorologically abnormal conditions, but also short-term plant water deficit degree [25]. The CI is widely used to monitor drought in China, and classifies drought status into five categories: non-drought for CI>-0.6; slight drought for −1.2 < CI ≤ −0.6; moderate drought for −1.8 < CI ≤ −1.2; severe drought for −2.4 < CI ≤ −1.8; and extreme drought for CI ≤ −2.4. The monthly mean CI values for all stations in Hubei Province were plotted in Figure 3.

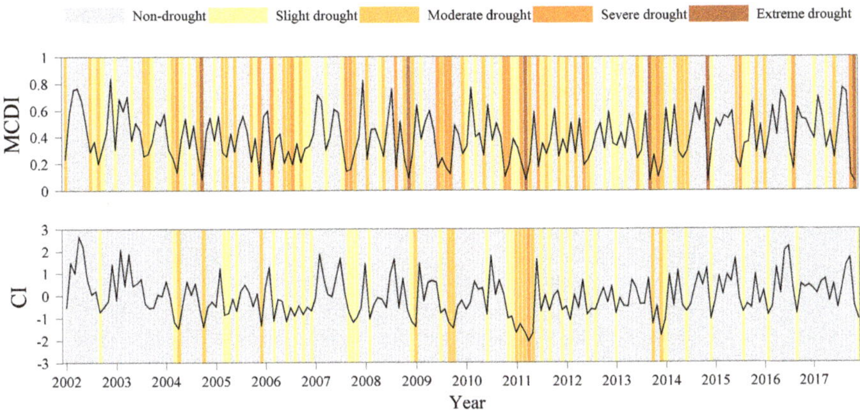

Figure 3. Monthly mean MDCI and CI values and the corresponding drought states for all stations in Hubei Province.

The MCDI and CI series showed similar trends from 2002–2017, and Pearson correlation analysis produced a very high coefficient of 0.89, indicating consistency in the assessment of dry condition variation by these two indices. However, differences in drought state classification between these indices were also clear from the figure. The drought states in several months (e.g., October 2004, December 2008, April 2011) were classified as extreme by MCDI, but moderate or severe by CI. To verify the accuracy of drought assessment by MCDI, the historical drought records were used to assess the performance. According to the "China Meteorological Disaster Record" [20], Hubei Province experienced a severe drought from November 2010 to May 2011, and the precipitation amount in most cities in Hubei Province was the lowest ever, with nearly 10 million people affected and 18.706 million mu of crops damaged. The first orange warning signal of meteorological drought in history was issued for this event. The figure showed that both indices accurately detected this drought event (onset and termination). However, the CI seemed to underestimate drought severity, with most months classified as slight drought, while MCDI precisely captured the drought onset with a value of 0.11 (severe drought) and labeled the driest month (April 2011) as extreme drought, in better agreement with the recorded conditions.

4.3. Temporal Analysis of Drought in Hubei Province Based on MCDI

The temporal distribution characteristics of drought in Hubei Province from 2002 to 2017 were analyzed based on monthly MCDI of selected meteorological stations. The evolution trends of drought from the perspective of overall and seasonal changes were investigated. Figure 3 showed that droughts occurred in Hubei Province throughout the year, although most severe and extreme droughts happened during the late summer and autumn (July–October), while spring was the wettest season with the fewest droughts. To assess the variation of drought categories in recent years, the numbers of drought (D_0–D_4) occurred in four-year intervals (i.e., 2002–2005, 2006–2009, 2010–2013, and 2014–2017) at each station were listed in Table 2.

Table 2. Number of Drought Category Events in Four-year intervals from 2002 to 2017.

Station	2002–2005					2006–2009					2010–2013					2014–2017				
	D_0	D_1	D_2	D_3	D_4	D_0	D_1	D_2	D_3	D_4	D_0	D_1	D_2	D_3	D_4	D_0	D_1	D_2	D_3	D_4
57251	22	8	8	7	3	24	3	4	9	8	16	5	9	11	7	18	3	16	7	4
57279	28	4	5	8	3	21	6	7	4	10	15	9	8	6	10	28	4	3	9	4
57285	20	5	7	10	6	19	4	8	9	8	12	9	9	4	14	26	8	3	6	5
57359	24	7	11	4	2	20	8	7	3	10	22	9	4	7	6	22	10	4	4	8
57363	23	7	10	6	2	27	6	1	10	4	17	6	11	8	6	25	5	5	4	9
57378	26	4	9	5	4	22	7	4	4	11	12	10	8	10	8	20	6	9	8	5
57395	19	8	13	6	2	18	7	6	7	10	11	8	12	9	8	20	8	5	6	9
57447	21	7	7	6	7	26	3	7	8	4	18	9	8	8	5	24	7	11	5	1
57458	24	8	6	6	4	24	6	8	6	4	23	8	6	7	4	27	5	8	3	5
57461	23	9	5	4	7	25	5	8	6	4	18	4	13	11	2	23	8	5	6	6
57476	22	5	2	12	7	20	8	8	5	7	14	7	14	9	4	25	9	3	5	6
57482	22	7	6	7	6	19	5	9	7	8	14	12	9	6	7	26	9	2	6	5
57494	22	9	9	3	5	18	5	11	10	4	17	14	5	5	7	22	12	2	5	7
57595	24	5	3	9	7	19	8	7	9	5	27	6	3	10	2	29	7	5	5	2
58402	20	6	6	9	7	20	9	7	6	6	19	9	5	5	10	27	7	5	4	5
58407	15	9	8	13	3	17	7	8	9	7	15	9	8	10	6	25	10	5	6	2

On average, 43.97% of months exhibited no drought, and the percentages for D_1 to D_4 were 14.87%, 14.58%, 14.39%, and 12.17%, respectively. The number of non-drought months did not show an obvious trend; however, the number of extreme drought (D_4) months greatly increased for most of the stations during the period 2006–2009. After that, the months of non-drought (D_0) at most stations decreased from 2010–2013 and then increased from 2014–2017. Overall, it seemed that the period from 2010 to 2013 was the driest, and then the drought frequency for D_1–D_4 became stable or gradually decreased. This may be due to the increased temperature and decrease in rainy days under global warming; however, intense precipitation became more frequent and heavier, leading to increases in evaporation and water content at the surface (as shown in previous Figure 2) to alleviate the dry conditions.

4.4. Spatial Analysis of Droughts in Hubei Province

To investigate the spatial distribution of droughts in Hubei Province based on MCDI monitoring, the seasonal drought frequency from 2002 to 2017 (i.e., total number of D_1–D_4 in spring (March–May), summer (June–August), autumn (September–November), and winter (December–February)) were plotted in Figure 4 to show drought-prone areas.

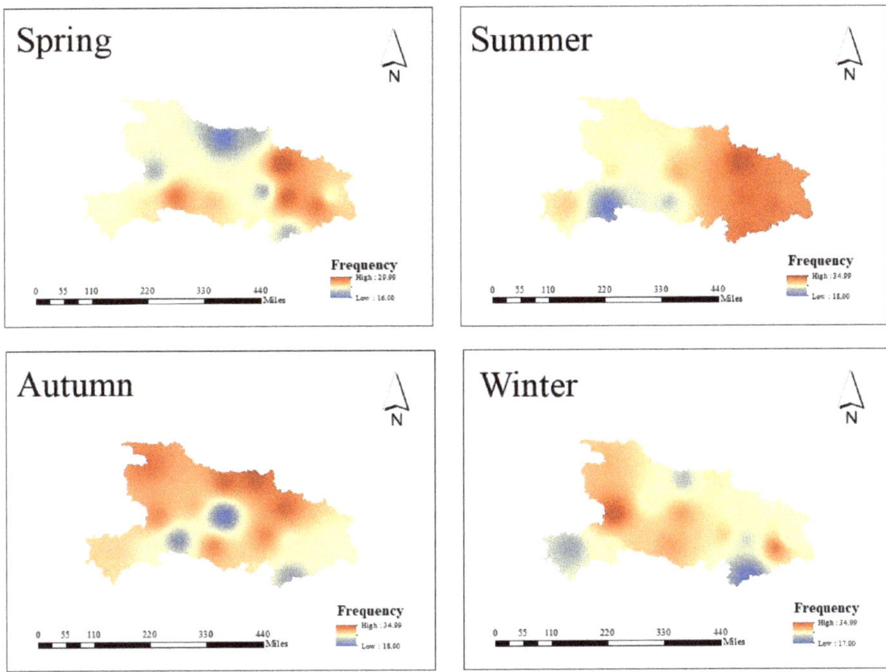

Figure 4. Spatial distribution of drought (D1-D4) frequency in each season for Hubei Province.

The figure showed spatial pattern differences of drought frequency in Hubei Province in different seasons. In spring and summer, the northeast region exhibited the largest drought frequency, while autumn and winter showed the most drought in the northwest and southwest regions, respectively. The northeast part of Hubei Province comprises hilly lands, with the majority of annual precipitation being concentrated in the monsoon, leading to more spring and early summer droughts in this area. The northwest part of Hubei Province is mostly mountainous regions located in the zone of the least rainfall and the most aridity, especially for the period after monsoon, leading to more autumn droughts. The southwest part is a mountainous region with sufficient rainfall throughout the year and exhibited sporadic late autumn and winter droughts. The central Jianghan Plain contains the watershed of the Yangzte River and the middle-lower reaches of the Han River, with more than 300 lakes, resulting in good water conservancy conditions and low frequency of droughts throughout the year.

Regardless of the seasonal differences in spatial characteristics, the annual frequency of drought categories D_2–D_4 (i.e., moderate, severe, and extreme), which result in serious drought events and damage to plants in Hubei Province, was plotted in Figure 5. The northwest and northeast regions were prone to serious droughts, while the central and southern parts of Hubei province were much less affected.

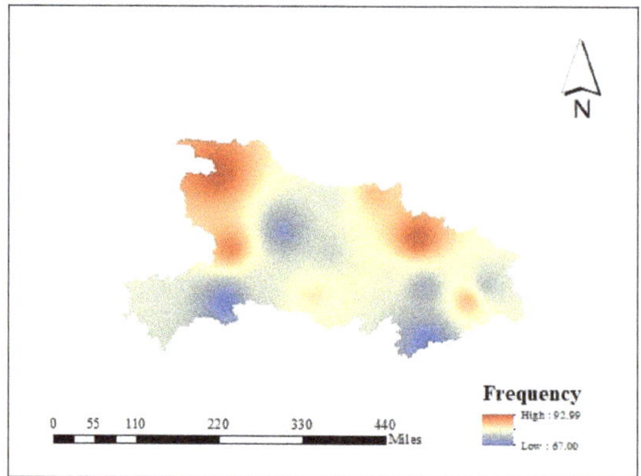

Figure 5. Spatial distribution of annual severe drought (D_2-D_4) frequency for Hubei Province.

5. Discussion

A large number of studies have reached a consensus that constructing drought indexes based on a single variable is likely to be insufficient for accurate drought risk assessment and reasonable decision-making [26]. Due to the complex physical interactions among drought indicators, (e.g., precipitation, evapotranspiration, infiltration, groundwater flow, etc.), the drought status acquired from one indicator often does not match well with that obtained from a different one. Thus, the investigation of drought structure based on an integrated drought index combining multivariate information deserves more attention. How to blend different drought-related variables by forming a latent drought variable/index through mathematical transformations is an area in which a number of researchers are conducting studies [27]. In this study, the MCDI was developed based on the information entropy to calculate the assigned weights for the inputs including precipitation, temperature, evaporation, and surface water content index on a monthly time scale to reflect the agrometeorological drought information. It is easy to implement the MCDI as a univariate drought index to display the drought conditions across different time scales with inputs aggregated at the corresponding time scales (such as 3-month time scale), and the potential limitation would be the adequacy of representativeness of hydrological phenomena. However, the proposed MCDI could be easily generalized by using additional input variables for considering local hydrological conditions, and the drought monitoring based on MCDI in Hubei Province in this study provided a good reference for other regions to conduct comprehensive drought assessment. Lastly, it is noted that the meteorological inputs for the MCDI were the observations from 16 meteorological stations across Hubei province instead of the remote-sensing products since that the ground information had certain advantages over remotely-sensed data. The ground observations are not only accurate, but can also reveal the influence of environmental, anthropic, and other factors on the drought development process. The continuity and stability of long-term meteorological observations constitute the basic dataset for comprehensive drought monitoring, except for the shortcomings in displaying drought spatial variability at lower resolutions. There were only 16 stations across Hubei Province to be applied, but these sites were mostly evenly distributed throughout the study area, covering the five areas of various topographic and climatic features (i.e., humid mountainous area in the southern part, the semi-humid plain area in the central part, the humid hilly area in the southeastern part, the arid mountainous area in the northwestern part, and the arid hilly area in the northern part) of Hubei Province, which were supposed to be able to reflect the spatial variability of drought conditions in the study area. By contrast, the remote sensing products have a wider geographical coverage and higher resolution, and are capable of monitoring the spatial characteristics of droughts for the ungauged regions or areas with poor observations. Thus, the MCDI could be further developed to capture drought conditions by using suitable data sources for selected drought indicators at different locations and time scales in the future research work.

6. Conclusions

In this study, a comprehensive drought monitoring index MCDI with precipitation, temperature, evaporation, and surface water content index (SWCI) as input variables was constructed. The index utilized observed meteorological data and remotely sensed SWCI data to provide comprehensive assessment of droughts in Hubei Province. Based on the results, several conclusions were reached. (1) The MCDI considers a combined effect of input variables, showing a similar tendency to the PA and a negative correlation with TA and ETA series; (2) Compared to the widely used meteorological drought composite index (CI), the MCDI had strong correlation but provided more accurate drought monitoring for historical drought events in Hubei Province. This indicates that the MCDI is a reliable comprehensive monitoring index to reflect composite information of meteorological and agricultural droughts; (3) Temporal drought analysis based on MCDI monitoring showed that droughts in Hubei Province may occur at any time in the year, but late summer and autumn were prone to severe and extreme droughts, and the spring season had the fewest droughts. On average, 43.97% of the year exhibited non-drought, and the percentages for drought categories D_1 to D_4 were nearly equal. The drought frequency increased during 2010–2013 and became stable or gradually decreased with the increased temperature and reduced rainy days; (4) Spatial drought frequency analysis based on the MCDI showed seasonal differences in distribution of drought-prone areas, with the northwest and northeast parts of Hubei province being prone to serious droughts, while the central and southern regions were much less affected by serious droughts; (5) Drought monitoring based on MCDI in Hubei Province provided a good reference for other regions to conduct comprehensive drought assessment and could be easily generalized using additional input variables for local conditions; (6) The MCDI in this study has limitations due to the short records of the remotely-sensed data products and the short period of drought monitoring, which will inevitably affect the accuracy of the results. Moreover, to obtain more reliable and robust results, meteorological stations in the study area should be located in various topographic regions.

Author Contributions: S.C.: Writing-original draft; W.Z.: Writing-original draft and investigation; S.P.: Validation; Q.X.: Conceptualization; T.-W.K.: Writing-review & editing and project administration. All authors have read and agreed to the published version of the manuscript.

Funding: This research was sponsored by the Natural Science Foundation of Hubei Province of China (2019CFB188), the Project (2018A003) of Hubei Key Laboratory of Regional Development and Environmental Response (Hubei University), the National Natural Science Foundation of China (41401186), and the Natural Science Foundation of Hubei Province of China (2019CFB538). Prof. Kim was supported by the Korea Environment Industry & Technology Institute (KEITI) through the Advanced Water Management Research Program, funded by the Korea Ministry of Environment (Grant.83070).

Conflicts of Interest: The authors declare no conflict of interest.

References

1. Wilhite, D.; Svoboda, M.; Hayes, M. Understanding the complex impacts of drought: A key to enhancing drought mitigation and preparedness. *Water Resour. Manag.* **2007**, *21*, 763–774. [CrossRef]
2. Prabhakar, S.; Shaw, R. Climate change adaptation implications for drought risk mitigation: A perspective for India. *Clim. Chang.* **2008**, *88*, 113–130. [CrossRef]
3. McKee, T.B.; Doesken, N.J.; Kleist, J. The relationship of drought frequency and duration to time scales. In Proceedings of the 8th Conference on Applied Climatology, Anaheim, CA, USA, 17–22 January 1993.
4. Palmer, W.C. *Meteorological Drought*; U.S. Department of Commerce: Washington, DC, USA, 1965.
5. Ju, X.S.; Yang, X.W.; Chen, L.J. Study on determination of drought and flood indices for single station and classification of regional drought and flood levels in China. *Q. J. Appl. Meteorol.* **1997**, *8*, 26–33. (In Chinese)
6. Wang, Z.W.; Qu, P.M. Analysis of drought characteristics in northern China in recent 50 years. *Acta Geogr. Sin.* **2003**, *58*, 61–68. (In Chinese)

7. Mo, K.C. Model-based drought indices over the United States. *J. Hydrometeorol.* **2008**, *9*, 1212–1230. (In Chinese) [CrossRef]
8. He, B.; Lü, A.; Wu, J.; Zhao, L.; Liu, M. Drought hazard assessment and spatial characteristics analysis in China. *J. Geogr. Sci.* **2011**, *21*, 235–249. (In Chinese) [CrossRef]
9. Waseem, M.; Ajmal, M.; Lee, J.H.; Kim, T.W. Multivariate drought assessment considering the antecedent drought conditions. *Water Resour. Manag.* **2016**, *30*, 4221–4231. [CrossRef]
10. Bonaccorso, B.; Cancelliere, A.; Rossi, G. Probabilistic forecasting of drought class transitions in Sicily (Italy) using Standardized Precipitation Index and North Atlantic Oscillation Index. *J. Hydrol.* **2015**, *526*, 136–150. [CrossRef]
11. Qian, L.L.; He, Z.H.; Liang, H.; Yang, Z.H.; Zeng, X.B. Spatial-temporal evolution characteristics of agricultural drought based on precipitation Z index in Guizhou. *J. Guizhou Norm. Univ.* **2019**, *37*, 10–14. (In Chinese)
12. Du, X.; Wang, S.X.; Zhou, Y. Construction and validation of a new model for Unified Surface Water Capacity based on EOS/MODIS data. *Geomat. Inf. Sci. Wuhan Univ.* **2007**, *32*, 205–208. (In Chinese)
13. Carlson, T.N.; Perry, E.M.; Schmugge, T.J. Remote sensing estimation of soil moisture availability and fractional vegetation cover for agricultural fields. *Agric. Meteorol.* **1990**, *52*, 45–69. [CrossRef]
14. Du, L.T.; Tian, Q.J.; Wang, L.; Huang, Y.; Nan, L. Construction of comprehensive drought monitoring model based on multi-source remote sensing data. *Trans. Chin. Soc. Agric. Eng.* **2014**, *30*, 126–132. (In Chinese)
15. Jiang, X.W.; Bai, J.J.; Liu, X.F. Research progress and prospect of integrated drought monitoring based on multi–source information. *Adv. Earth Sci.* **2019**, *34*, 275–287. (In Chinese)
16. Hao, Z.C.; AghaKouchak, A. Multivariate standardized drought index: A parametric multi-index model. *Adv. Water Resour.* **2013**, *57*, 12–18. [CrossRef]
17. Kao, S.C.; Govindaraju, R.S. A copula-based joint deficit index for droughts. *J. Hydrol.* **2010**, *380*, 121–134. [CrossRef]
18. Sun, L.; Mitchell, S.W.; Davidson, A. Multiple drought indices for agriculture drought risk assessment on the Canadian prairies. *Int. J. Climatol.* **2012**, *32*, 1628–1639. [CrossRef]
19. Hao, C.; Zhang, J.; Yao, F. Combination of multi-sensor remote sensing data for drought monitoring over Southwest China. *Int. J. Appl. Earth Obs. Geo Inf.* **2015**, *35*, 270–283. (In Chinese) [CrossRef]
20. China National Meteorological Information Center. *China Natural Disaster Statistics Yearbook*; China Statistics Press: Beijing, China, 2018.
21. Wang, Z.D.; Guo, P.; Wan, H.; Yang, G. Analysis of drought monitoring in Shandong Province from 2014 to 2016 based on MODIS data. *Res. Soil Water Conserv.* **2019**, *2*, 330–336. (In Chinese)
22. Yu, M.; Cheng, M.H. Drought monitoring in Heilongjiang Province based on NDVI-Ts space. *J. Appl. Meteorol. Sci.* **2010**, *21*, 221–228. (In Chinese)
23. Waseem, M.; Ajmal, M.; Kim, T.W. Development of a new composite drought index for multivariate drought assessment. *J. Hydrol.* **2015**, *527*, 30–37. [CrossRef]
24. Shannon, C. A Mathematical Theory of Communications. *Bell Syst. Technol. J.* **1948**, *27*, 379–423. [CrossRef]
25. Li, S.Y.; Liu, R.H.; Shi, L.K.; Ma, Z.H. Analysis on drought characteristic of Henan in recent 40 years based on Meteorological Drought Composite Index. *J. Arid Meteorol.* **2009**, *2*, 97–102. (In Chinese)
26. Huang, S.Z.; Huang, Q.; Chang, J.X.; Zhu, Y.L.; Leng, G.Y.; Xing, L. Drought structure based on a nonparametric multivariate standardized drought index across the Yellow River basin, China. *J. Hydrol.* **2015**, *530*, 127–136. [CrossRef]
27. Hao, Z.C.; Singh, V.P. Drought characterization from a multivariate perspective: A review. *J. Hydrol.* **2015**, *527*, 668–678. [CrossRef]

Article

Effect of Climate Change on Maize Yield in the Growing Season: A Case Study of the Songliao Plain Maize Belt

Ari Guna [1,2,3], Jiquan Zhang [1,2,3,*], Siqin Tong [1,2], Yongbin Bao [1], Aru Han [1] and Kaiwei Li [1]

1   School of Environment, Northeast Normal University, No.2555 Jingyue Street, Nanguan, Changchun 130117, China; argn078@nenu.edu.cn (A.G.); tongsq223@nenu.edu.cn (S.T.); baoyb924@nenu.edu.cn (Y.B.); arh690@nenu.edu.cn (A.H.); likw395@nenu.edu.cn (K.L.)
2   State Environmental Protection Key Laboratory of Wetland Ecology and Vegetation Restoration, Northeast Normal University, No.2555 Jingyue Street, Nanguan, Changchun 130117, China
3   Key Laboratory for Vegetation Ecology, Ministry of Education, No.2555 Jingyue Street, Nanguan, Changchun 130117, China
*   Correspondence: zhangjq022@nenu.edu.cn; Tel.: +86-0431-8916-5616

Received: 31 August 2019; Accepted: 6 October 2019; Published: 10 October 2019

Abstract: Based on the 1965–2017 climate data of 18 meteorological stations in the Songliao Plain maize belt, the Coupled Model Intercomparision Project (CMIP5) data, and the 1998–2017 maize yield data, the drought change characteristics in the study area were analyzed by using the standardized precipitation evapotranspiration index (SPEI) and the Mann–Kendall mutation test; furthermore, the relationship between meteorological factors, drought index, and maize climate yield was determined. Finally, the maize climate yields under 1.5 °C and 2.0 °C global warming scenarios were predicted. The results revealed that: (1) from 1965 to 2017, the study area experienced increasing temperature, decreasing precipitation, and intensifying drought trends; (2) the yield of the study area showed a downward trend from 1998 to 2017. Furthermore, the climate yield was negatively correlated with temperature, positively correlated with precipitation, and positively correlated with SPEI-1 and SPEI-3; and (3) under the 1.5 °C and the 2.0 °C global warming scenarios, the temperature and the precipitation increased in the maize growing season. Furthermore, under the studied global warming scenarios, the yield changes predicted by multiple regression were −7.7% and −15.9%, respectively, and the yield changes predicted by one-variable regression were −12.2% and −21.8%, respectively.

Keywords: drought; global warming; maize yield; Songliao Plain maize belt

---

1. Introduction

The agriculture sector is closely linked to sustainability and food security, as at least 159 million people, 2.19% of the world population, are directly or indirectly dependent on crop production; however, the agriculture sector is extremely vulnerable to climate-induced risks and natural disasters [1]. The Fourth Assessment Report of the Intergovernmental Panel on Climate Change (IPCC.AR4) [2] points out that the earth has experienced significant changes characterized by global warming in the past 100 years. To avoid possible harm to the ecological environment caused by global warming, The Paris Agreement explicitly proposes controlling the global temperature rise within 2.0 °C and within 1.5 °C above the industrial level [3]. Unstable global climate change will lead to an increase or a decrease in the instability of agricultural production [4,5]. Drought formation is due not only to numerous natural environments but also to human interventions. In the field of agriculture, inappropriate temperature, precipitation, humidity, and other conditions in the key growth period of crops eventually lead to yield changes. The abundance of grain production will directly affect national food security. Maize is the most important crop in northeast China. Drought is mainly distributed in northeast, north, and northwest China [6]. Owing to the impact of drought disasters, the production and the price of foreign maize trade circulation and other fields have undergone significant changes. The risk of unstable maize production caused by drought persists; it will directly affect the world maize trade circulation pattern and poses a serious threat to global food security. Therefore, it is necessary to explore and evaluate climate change and drought disasters and further study their impact on crop yield. Based on the literature at home and abroad, the impacts of climate variability and change on crop yields have been studied by numerous researchers worldwide both for historical and future climates and for various crops [7,8].

Presently, there are three main methods for studying the impact of climate change on grain crop yield: the field test method, which can be used to find the correlation between climate change characteristics and farmland crop management practices and the impact of climate change on agricultural production, the global climate model (GCM), and the regional climate model (RegCM), as well as other predictions of different climate change scenarios on crop yield [4] or by adding meteorological variables to the production function model [9]. Among them, Coupled Model Intercomparision Project (CMIP5) introduces the earth system model and its biogeochemical process, enabling realization of the global carbon cycle process and the dynamic preparation process, the use of statistical data to establish a correlation matrix between meteorological factors and crop yield, quantitative analysis of crop yield to climate change law, and response [7]. For example, some studies reported reductions in wheat yields in France and maize yields in China as a result of increased temperature [10,11]. In the United States, the yield impacts on maize, wheat, and soybean are not obvious owing to the less significant climate trends [12]. Gupta et al. [13] used the method of combining the GCM and a crop model to predict the change in rice yield in India. Chen et al. [14] calculated the linear trend of time and found that an increase in temperature increased the national single-season rice yield by 11%.

Future climate change research on crop yield focuses on the climate change background of global temperature rises of 1.5 °C and 2.0 °C. Huang et al. [15] showed that the global warming in arid and semi-arid regions has been 20–40% higher than in humid regions in the past 100 years. When the average global warming in the future reaches 2.0 °C, climate disasters such as maize yield reduction, surface runoff reduction, drought intensification, and malaria transmission caused by an increase in temperature will be the most serious in arid and semi-arid regions. Under the background of global warming, the yields of rice and maize have been reduced by 10% and 3.8%, respectively [16,17]. Chen et al. [18] mainly discussed the distribution characteristics of the main crop yields in China and analyzed the spatial variation characteristics of flowering period, maturity period, yield, evapotranspiration (ET), and water use efficiency (WUE) of maize in China's maize planting areas under the 1.5 °C and the 2.0 °C global warming scenarios .

The above research is mainly based on meteorological factors. In fact, as drought indices are related to the cumulative effects of a prolonged and abnormal moisture deficiency, they have a strong connection to agriculture. A number of studies related to drought analysis and monitoring have been conducted using the precipitation anomaly percentage (Pa) [19], the Palmer drought severity index (PDSI) [20], the standardized precipitation index (SPI) [21], the integrated meteorological drought index (CI) [22], the standardized precipitation evapotranspiration index (SPEI) [23], etc. For example, Zhang et al. [24], using the SPEI to describe drought situations and evaluate the relationship between drought risk and yield in each growth period, found that drought disaster in the entire growth period is closely related to maize yield. Matiu et al. [25] studied the relationship between crop yield and growth season temperature and SPEI in major maize, rice, soybean, and wheat producing countries and found that the combined effects of high temperature and drought significantly reduced the yield of maize, soybean, and wheat by 11.6%, 12.4%, and 9.2%, respectively. The SPEI combines the advantages of SPI and PDSI, taking into account the effect of evaporation on drought. It is sensitive to temperature changes and has the advantage of allowing for a convenient SPI calculation. It is suitable for multi-scale and multi-space comparison and can more sensitively reflect drought characteristics under the background of climate warming. In designing the SPEI computation, Vicente-Serrano et al. [23] followed the same theoretical basis that McKee et al. [21] devised to improve the SPI; moreover, the SPEI also has good applicability in China.

The Songliao Plain maize belt is the main body of the northeast industrial base, shouldering the great responsibility of food security in the country, and is also a traditional agricultural area dominated by food production. In recent decades, the climate change in the maize belt of the Songliao Plain has been remarkable, which has had an obvious impact on the maize production in this area. Fifty-five percent of the grain output reduction in the central Songliao Plain is caused by agro-meteorological disasters. Drought, which accounts for 60% of agro-meteorological disasters, occurs frequently in the central part of Songliao Plain [26]. Therefore, the aim of this study was to predict the change trend of maize yield under the 1.5 °C and the 2.0 °C global warming scenarios through regression equations obtained from historical climate change data and maize yield. Specifically, we (1) studied the climate change characteristics during the growing season from 1965 to 2017, i.e., temperature, precipitation, and drought characteristics SPEI, (2) established their relationship with maize yield, and (3) finally predicted the future climate change trend and maize yield change in the Songliao Plain maize belt under the 1.5 °C and the 2.0 °C global warming scenarios.

2. Materials and Methods

2.1. Study Area and Data

The Songliao Plain maize belt is located in northeast China at 40°12′–46°18′ N, 120°42′–127°36′ E, including northeast Inner Mongolia, southern Heilongjiang Province, central and western Jilin Province, and most of Liaoning province. The entire area is divided into 17 prefectural regions. The maize belt in the Songliao Plain is surrounded by piedmont diluvial, alluvial plain, and terrace. The plain is flat. The Songliao watershed divides the plain into two parts. North of the watershed is the Songnen Plain, which was formed by flooding of the Songhua River and the Nenjiang River. The middle and the lower reaches of the Liaohe River south of the watershed are called the Liaohe Plain (Figure 1). These provide excellent conditions for maize cultivation in this area. The Songliao Plain maize belt is located in a temperate zone and warm temperate zone with continental and monsoon climate characteristics. Summer is hot and rainy, while winter is cold and dry. Due to the influence of the monsoon climate, 60–65% of the precipitation in the crop growing season from May to September is concentrated in summer. The average temperature in the growing period for maize is 19–25 °C, the accumulated temperature ≥10 °C is 2800–3200 °C, the precipitation is 300–750 mm, and the frost-free period is 135–155 day.

The meteorological data used in this research were obtained from the National Meteorological Information Center (http://www.resdc.cn/), including monthly mean temperature and precipitation data from 1965 to 2017 at 18 meteorological stations. The yield data were obtained from statistical yearbooks of the Liaoning, Jilin, Heilongjiang, and Inner Mongolia provinces, including the maize yield of each prefecture-level city from 1998 to 2017. The trend yield and the climate yield were calculated by the linear moving average method. The CMIP5 climate model data provided by the Inter-Sectional Impact Model Inter-Comparison Project (ISI-MIP, http:www.isi-mip.org) were used in this study. These data include five climate models from 1850 to 2100, GFDL-ESM2M, IPSL-CM5A-LR, NorESM1-M, MIROC-ESM, and HadGEM2-ES, temperature and precipitation data under four future representative concentration pathways (RCP) scenarios (RCP2.6, RCP4.5, RCP6.0, and RCP8.5), and historical data. Taking 1986–2005 as the reference period, owing to the different resolutions of global climate models, the data of different resolutions were uniformly interpolated to a grid of 0.5° × 0.5° by bilinear interpolation.

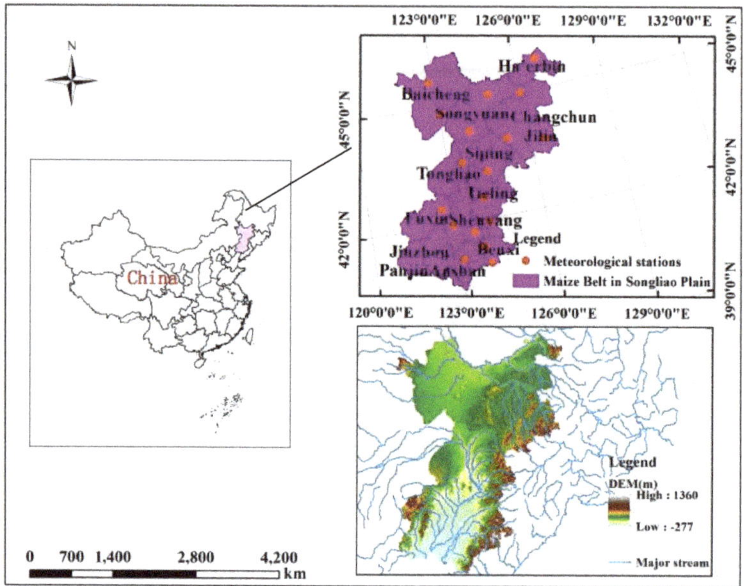

Figure 1. Study area location, meteorological stations, and major river network.

2.2. Methodology

2.2.1. Standardized Precipitation Evapotranspiration Index

To monitor and study the drought process under the background of global warming, Vicente-Serrano et al. [27] proposed the SPEI. The SPEI index is an ideal index for drought monitoring and assessment under the background of climate warming and is widely used in drought research [28,29]. The specific calculation process is shown in reference [23]. On the basis of referring to the drought conditions recorded in the "China Meteorological Drought Atlas (1956–2009)" over the years in the research area and based on the National Meteorological Drought Grade Standard (GB/T20481-2017) implemented in 2017, this study formulated an SPEI drought grade classification standard (Table 1) suitable for the research area.

Table 1. Classification standard for drought based on the standardized precipitation evapotranspiration index (SPEI).

Grade	Type	SPEI Value
1	Normal	$-0.5 < \text{SPEI}$
2	Light drought	$-1.0 < \text{SPEI} \leq -0.5$
3	Moderate drought	$-1.5 < \text{SPEI} \leq -1.0$
4	Severe drought	$-2.0 < \text{SPEI} \leq -1.5$
5	Extreme drought	$\text{SPEI} \leq -2.0$

2.2.2. Mann–Kendall Mutation Test

The Mann–Kendall (M–K) mutation test was originally proposed and developed by H.B. Mann [30] and M. G. Kendall [31]. The Mann–Kendall mutation test method is a nonparametric method. It has sample sequences that do not need to follow a certain distribution and are not affected by a few outliers. It is suitable for sequence variables and type variables [32]. The calculation principle of the Mann–Kendall mutation test is as follows.

Assuming that the time series x_1, x_2, ..., x_n, S_k represents the cumulative number of sample $X_i > X_j$ $(1 \leq i \leq j)$ in the I sample, define statistics:

$$S_k = \sum_{i=1}^{k} r_i, \; r_i = \left\{ \begin{array}{l} 1, \; x_i < x_j \\ 0, \; x_i \geq x_j \end{array} \right. , \; j = 1, 2, \dots; k = 1, 2, \dots, n \tag{1}$$

Under the assumption that the time series are randomly independent, the mean and the variance of S_k are:

$$[S_k] = \frac{K(K-1)}{4} \tag{2}$$

$$Var[S_k] = \frac{K(K-1)(2K+5)}{72} \; (a \leq k \leq n) \tag{3}$$

Standardizing S_k,

$$UF_k = \left(S_k - \frac{E[S_k]}{\sqrt{Var[S_k]}} \right) \tag{4}$$

where $UF_1 = 0$. Given the significance level α, if $|UF_k| > U_\alpha$, it indicates that the sequence has obvious trend changes, and all UF_k can form a curve. This method is applied to an inverse sequence, and the UB_k curve is calculated. Given the significance level, such as $\alpha = 0.05$, the critical value $U_{0.05} = \pm 1.96$. If the value of UF_k or UB_k is less than 0, the sequence shows a downward trend, and if it is greater than 0, the sequence shows an upward trend. When they exceed the critical line, the downward or the upward trend is significant. The range exceeding the critical line is determined as the time zone where the mutation occurs. If two curves UF_k and UB_k have intersection points, and the intersection points are between critical lines, the time corresponding to the intersection points is the time when the mutation starts.

2.2.3. Calculation of Climate Yield and Drought Assessment Indicators

The influencing factors of agricultural production are mainly divided into climatic factors and social science and technology factors [33]. To study the effect of drought on crop yield, it is necessary to obtain the climate yield. By identifying the long-term trend of yield unit area and using the appropriate de-trend method, the influence of climate on crop yield unit area can be obtained, i.e., the climate yield can be obtained [34].

$$Y_c = Y - Y_t \tag{5}$$

where Y is the actual unit yield, Y_t is the trend yield, and Y_c is the climate yield, which is the fluctuating yield component affected by climatic factors and can reflect favorable or unfavorable climatic conditions and their effects on the yield year by year.

In this study, the drought frequency was selected as the evaluation index [35]. The drought frequency refers to the ratio of the number of droughts of a certain grade to the total number of years in a statistical period of a certain station. The calculation formula is as follows:

$$DF = \frac{n}{N} \times 100\% \tag{6}$$

where DF is the frequency of drought occurrence, N is the total length of the time series, n is the number of years that SPEI meets a certain level of drought intensity in the statistical period, and the larger DF indicates that drought occurs more frequently.

2.2.4. Correlation Analysis

Correlation analysis is a method used to describe the linear correlation of two variables. If there are two data sequences $X = \{x_1, x_2, \ldots \ldots, x_n\}$ and $Y = \{y_1, y_2, \ldots \ldots, y_n\}$, the correlation coefficient formula is calculated as follows:

$$r = \frac{cov(x,y)}{S_x S_y} = \frac{\sum_{i=1}^{n}(x_i - \bar{x})(y_i - \bar{y})}{\sqrt{\sum_{i=1}^{n}(x_i - \bar{x})^2}\sqrt{\sum_{i=1}^{n}(y_i - \bar{y})^2}} \tag{7}$$

where S_x and S_y are the standard deviations of variables X and y, respectively, and $cov(x,y)$ is the covariance of variables x and y. The correlation coefficient r ranges from -1.0 to 1.0. When $r > 0$, the two variables are positively correlated, when $r < 0$, the two variables are negatively correlated. The larger $|r|$ is, the stronger the correlation between the two variables is.

3. Results

3.1. Analysis of Climate Trend and Drought Index Changes

3.1.1. Analysis of the Variation Characteristics of Temperature and Precipitation

Figure 2a displays the average temperature change trend and the M–K mutation test during the growing season in the Songliao Plain maize belt. It can be seen that the temperature in the maize belt of the Songliao Plain shows an overall warming trend in the growing season, with a maximum temperature of 21.65 °C (2000) and a minimum temperature of 18.64 °C (1976) and therefore an average temperature of 19.94 °C. The UF and the UB curves crossed around 1995, indicating that a sudden temperature change occurred in 1995. Before this, the increasing temperature trend was not obvious, and after that, there was an obvious warming trend, with the significance level reaching 0.05. As shown in Figure 2b, the precipitation in the maize belt of the Songliao Plain has a downward trend from 1965 to 2017, but the change trend is not obvious, with maximum rainfall of 703.97 mm in 1994 and minimum rainfall of 318.37 mm in 2000 and an average of 481.18 mm. The UF and the UB curves frequently crossed before 1985 and crossed around 1995 and 2010 after 1985, which indicates that this is the mutation point of the precipitation change trend in the past 20 years. Moreover, there was a decreasing trend of precipitation after the abrupt change in 1995, but it was not significant.

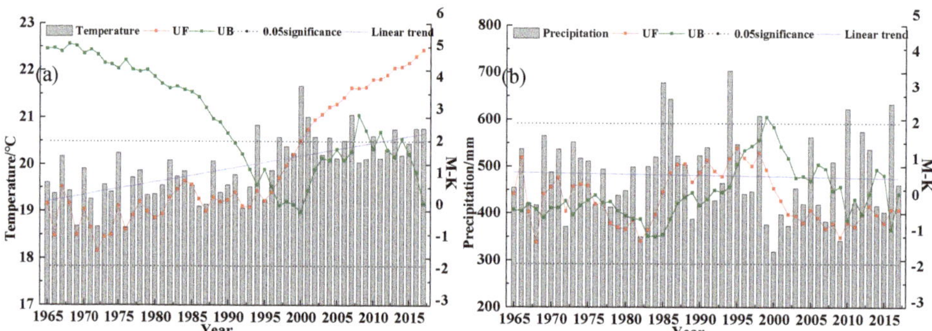

Figure 2. Variation and linear trend of temperature (**a**) and precipitation (**b**) in growing season and Mann–Kendall (M–K) mutation test.

As shown in Figure 3a, there is an obvious warming trend in all parts of the Songliao Plain maize belt. The average warming temperature of the whole plain from 1965 to 2017 is 0.28 °C/10a. The warming in most areas is between 0.25 and 0.35 °C/10a. The warming trend is the largest in Liaoyang, southern Songliao Plain maize belt (0.42 °C/10a), and the smallest in Fushun (0.15 °C/10a). Figure 3b is a graph showing the spatial variation trend of precipitation in the growing season. The spatial distribution of precipitation is uneven. A total of 60% of the stations in the maize belt in the Songliao Plain have a decreasing precipitation trend. These areas are concentrated in the south and the northwest of the maize belt in the Songliao Plain, while the northeast of the maize belt in the Songliao Plain has a significant increasing trend of 0–9.4 mm/10a.

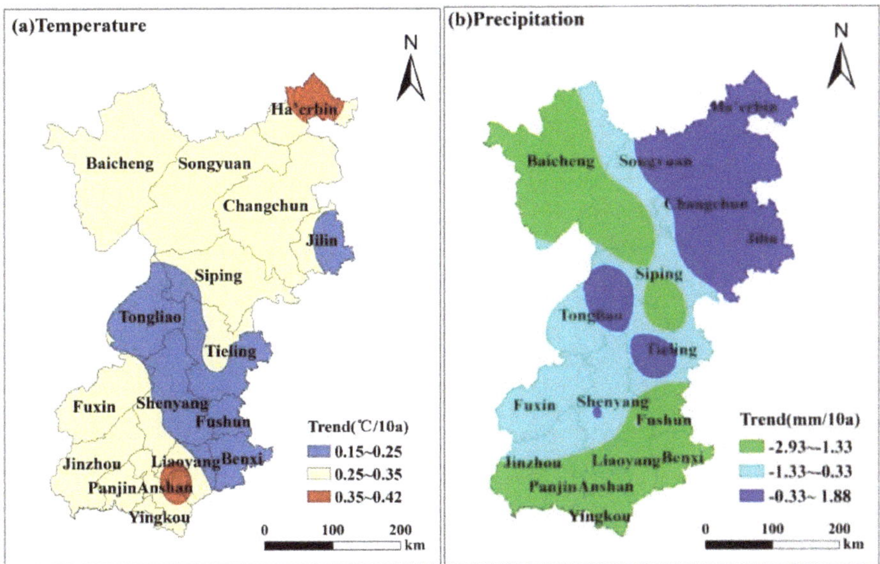

Figure 3. Spatial variation trend of temperature (**a**) precipitation (**b**) in the growing season during the period 1965–2017.

3.1.2. Analysis of Drought Change Characteristics

The SPEI can describe the succession of drought and flood events at different time scales. Furthermore, in combination with different grades of drought division (Table 1), it can evaluate the characteristics of different types of drought changes at different time scales in the study area. As can be seen from Figure 4, drought and flood events alternate frequently in a short time scale of one and three months, and SPEI-1 and SPEI-3 are very sensitive to changes in short-term temperature and precipitation (Figure 4a,b). With the prolongation of the SPEI time scale, the alternate frequency of drought and flood events decreases, showing a longer periodicity. SPEI-6 and SPEI-12 have weak responses to short-term temperature and precipitation, and drought changes are relatively stable. Figure 4d shows that maize in the Songliao Plain experiences alternating periods of drought and flooding of about two years, such as 1965–1966 and 1970–1971, with two drought periods of about six years occurring in 1985–1990 and 1998–2003.

The maize growing season generally lasts from May to September. As the September SPEI-6 is calculated by using the temperature and the precipitation sequence of the six months from April to September, which covers the entire growing season, the September SPEI-6 is used to describe the drought situation of the entire growing season. Table 1 is used as the standard for statistical analysis. Figure 5 is a spatial distribution diagram of the frequencies of light, moderate, severe, and extreme drought in the maize belt of the Songliao Plain. The frequency of occurrence from high to low is light drought > moderate drought > severe drought > extreme drought. The frequency of light drought (Figure 5a) is 7.5–24.5%, with Baicheng and Tieling having the highest frequency and Benxi having the lowest frequency; the frequency of moderate drought (Figure 5b) is 3.7–17%, and Tieling has the lowest frequency, while Shenyang has the highest frequency; the frequency of severe drought (Figure 5c) is 1.8–9.8%, and Jilin has the lowest frequency, while Songyuan, Fuxin, Anshan, Yingkou, and other regions have relatively high frequencies; the frequency of extreme drought (Figure 5d) is 0–4%, and Fushun has the highest frequency. Generally speaking, the higher the intensity of drought is, the lower its frequency is.

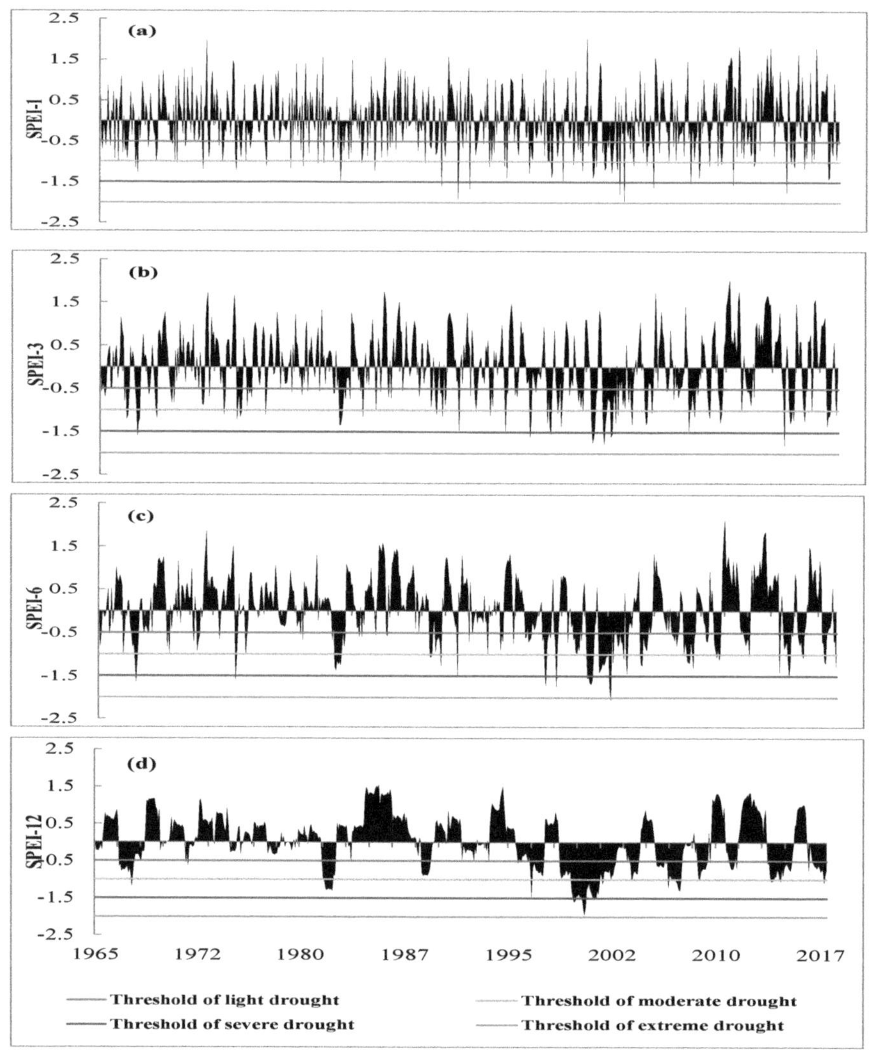

Figure 4. Time series variability of SPEI-1 (**a**); SPEI-3 (**b**); SPEI-6 (**c**); and SPEI-12 (**d**) in the time period 1965–2017.

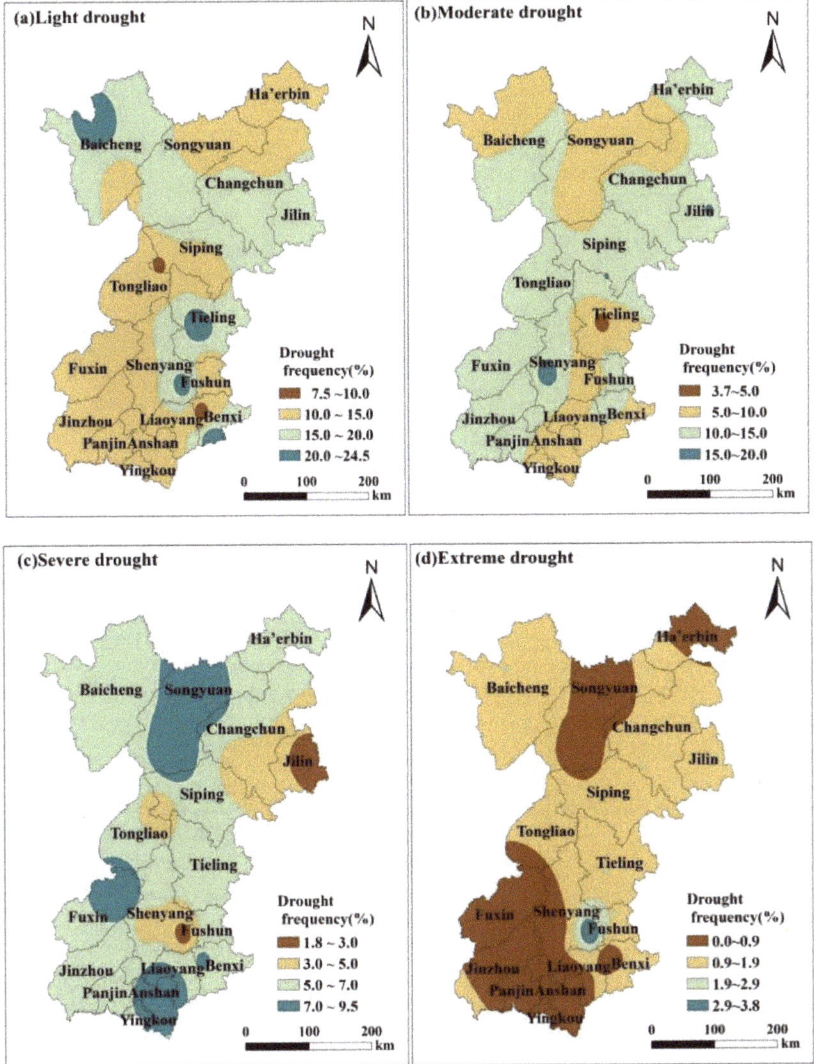

Figure 5. Spatial distribution of the frequencies of light drought (**a**); moderate drought (**b**); severe drought (**c**); and extreme drought (**d**) during the period 1965–2017.

3.2. Analysis of the Relationship between Climate Trend, Drought Index, and Yield

3.2.1. Analysis of the Maize Yield Changes in the Songliao Plain Maize Belt

Figure 6a shows the change in the maize yield unit area from 1998 to 2017 in the Songliao Plain maize belt. Of the 17 prefectural regions in the past 20 years, except Baicheng, Fuxin, Jinzhou, Panjin, Shenyang, and Tongliao, there is a trend of production reduction in the other 11 regions. Siping has the largest reduction trend 249 (kg/ha)/10a, followed by Jilin. In 2000, minimum production was achieved in all regions and reached 4929 kg/ha in the Songliao Plain. The highest yield of the maize belt in the Songliao Plain was 7643 kg/ha in 2008, and the maize yield tended to be stable after 2010. As can be seen from Figure 6b, the yield distribution in different areas of the Songliao Plain maize belt was also different. The average yield of the maize belt in the Songliao Plain is about 6808 kg/ha. The unit yield in Siping is 9811 kg/ha, which is twice as high as 4464 kg/ha in Tongliao. The unit yield of the central and the northeast regions (Harbin, Songyuan, Changchun, Siping, and Tieling) is higher than that of the northwest and the southeast regions.

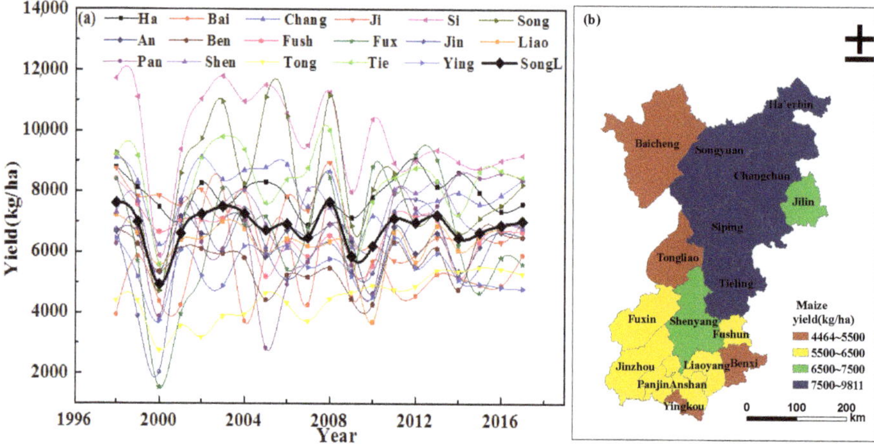

Figure 6. (**a**) Fluctuation in maize yield and spatial distribution (**b**) of maize yield in prefectural regions in the Songliao Plain maize belt during the period 1998–2017. Note: Ha'erbin: Ha, Baicheng: Bai, Changchun: Chang, Jilin: Ji, Siping: Si, Songyuan: Song, Anshan: An, Benxi: Ben, Fushun: Fush, Fuxin: Fux, Jinzhou: Jin, Liaoyang: Liao, Panjin: Pan, Shenyang: Shen, Tongliao: Tong, Tieling: Tie, Yingkou: Ying, Songliao: SongL.

3.2.2. Relationship between Temperature, Precipitation, and Yield

The calculation results (Table 2) show that the correlation between the climate yield of maize and the temperature in the corresponding growing season (May–September) has varied over the past 20 years in the maize belt of the Songliao Plain, indicating that there are regional differences in temperature sensitivity to different months. Apart from Jilin, Benxi, and Fushun, the temperature in 84% of the other regions is basically negatively related to the climate yield. Songyuan and Siping have the highest correlations of −0.71 and −0.73, respectively, ($P \leq 0.01$), which may be due to the higher warming amplitude in the growth season in these two regions. The correlation between the maize belt and the climate yield in the Songliao Plain is −0.67 ($P \leq 0.01$). All the months of the growing season except May have higher correlations with climate yield. However, relative to each month of the growing season, the mean value (May–September) has a higher correlation with the climate yield. This shows that the mean temperature value in the analysis growth season is a better reflection of the climate yield than the temperature value in each month.

Table 3 shows the correlation between climate yield and precipitation in various regions of the Songliao Plain maize belt in the last 20 years. Except for May, the correlation between climate yield and precipitation is basically positive, but the correlation is lower than that of temperature as a whole. July, the growth season, has the highest relationship with the climatic output of various regions, followed by June. The correlation between the climate yield and the summation of monthly precipitation in the growing season in Siping, Songyuan, and Fuxin is high, i.e., 0.58, 0.57, and 0.71, respectively ($P \leq 0.01$), which may be caused by a higher precipitation reduction. The summation of monthly precipitation in the growing season of the maize belt in the Songliao Plain has a significant correlation with climate yield, which is 0.51 ($P \leq 0.01$).

Table 2. Pearson's correlation coefficient of temperature and climate yield in maize growing season in Songliao Plain maize belt during the period 1998–2017.

Month	Ha	Bai	Chang	Ji	Si	Song	An	Ben	Fush	Fux	Jin	liao	Pan	Shen	Tong	Tie	Ying	SongL
May	−0.20	−0.11	−0.26	0.07	−0.37	−0.26	0.21	0.18	0.22	−0.20	0.16	0.13	0.14	−0.12	−0.26	−0.25	−0.10	−0.13
Jun.	−0.30	−0.44 *	−0.28	0.01	−0.35	−0.52 **	−0.39	−0.20	−0.17	−0.21	−0.32	−0.41 *	−0.16	−0.24	−0.29	−0.30	−0.30	0.50 *
Jul.	−0.02	−0.14	−0.29	−0.08	−0.41 *	−0.37	−0.23	0.20	0.04	−0.20	−0.27	−0.12	0.16	−0.16	0.27	−0.15	−0.20	−0.23
Aug.	−0.20	−0.18	−0.35	−0.24	−0.51 *	−0.26	−0.12	−0.25	−0.38	−0.41 *	−0.45 *	−0.01	−0.36	−0.31	−0.06	−0.43 *	−0.02	−0.47 *
Sep.	−0.22	−0.20	0.19	0.15	−0.08	−0.28	−0.25	0.23	0.11	−0.13	−0.12	−0.03	−0.12	−0.15	−0.07	0.09	−0.33	−0.12
mean	−0.33	−0.48 *	−0.45 *	−0.02	−0.71 **	−0.73 **	−0.32	0.04	−0.05	−0.45 *	−0.37	−0.23	−0.10	−0.39	−0.23	−0.45 *	−0.40	−0.67 **

Note: * Significant at $P \leq 0.05$; ** significant at $P \leq 0.01$; red and blue indicate positive correlation and negative correlation respectively.

Table 3. Pearson's correlation coefficient of precipitation and climate yield in the maize growing season in Songliao Plain maize belt during the period 1998–2017.

Month	Ha	Bai	Chang	Ji	Si	Song	An	Ben	Fush	Fux	Jin	liao	Pan	Shen	Tong	Tie	Ying	SongL
May	−0.09	0.04	0.04	−0.23	0.13	0.14	−0.29	−0.24	−0.41	0.18	−0.02	−0.30	−0.12	−0.11	0.35	0.08	−0.10	−0.03
Jun.	0.53 *	0.26	0.29	−0.21	0.32	0.42	0.13	−0.10	0.06	0.28	0.37	0.14	0.09	0.23	0.12	0.14	0.30	0.34
Jul.	0.01	0.15	0.33	0.10	0.57*	0.49	0.07	0.07	0.30	0.60 *	0.43 *	−0.14	−0.15	0.25	0.06	0.41 *	0.14	0.41 *
Aug.	0.13	0.02	0.19	−0.08	0.35	0.33	−0.13	0.09	0.22	0.53 *	0.30	0.18	−0.26	0.01	0.30	0.26	0.17	0.26
Sep.	0.17	−0.04	0.13	0.32	0.16	0.14	0.03	−0.06	0.05	0.48 *	0.36	0.03	0.07	0.22	0.40 *	0.22	0.13	0.44 *
sum	0.24	0.16	0.35	−0.20	0.58 **	0.57 **	−0.07	−0.06	0.11	0.71 **	0.52 *	−0.17	0.17	0.29	0.39	0.39	0.32	0.51 **

Note: * Significant at $P \leq 0.05$; ** significant at $P \leq 0.01$; red and blue indicate positive correlation and negative correlation respectively.

The results of the regression analysis of the temperature, the precipitation, and the climate yield in the maize belt of the Songliao Plain reveals that the two are in line with those of the multiple linear regression. Among them, the goodness-of-fit (R^2) of the regression equations established for the growth season average temperature, precipitation, and maize yield in 58% of the areas of the Songliao Plain maize belt is above 0.45, with the highest significance ($P \leq 0.01$) being exhibited in Siping ($R^2 = 0.66$), followed by the Songliao Plain maize belt ($R^2 = 0.54$). The linear regression equation form for the Songliao Plain maize belt is:

$$y = 1.916x_1 - 782.566x_2 + 15174.997 \tag{8}$$

where y is the maize climate yield, x_1 is the precipitation in the growing season, and x_2 is the average temperature in the growing season. From the regression equation, it can be seen that the maize climate yield in the Songliao Plain maize belt is positively correlated with the precipitation in the growing season and negatively correlated with the average temperature in the growing season.

3.2.3. Relationship between the SPEI of Different Scales and Yields

According to previous studies [36], the results of a correlation analysis between the climate yield and the SPEI can reflect the sensitivity of the yield to dry and wet changes at different time scales. The correlation between the climate yield and the SPEI on a short time scale is high, thus SPEI-1 and SPEI-3 are selected for further analysis. The correlation analysis results between the climate yield and the SPEI-1 in the Songliao Plain maize belt are shown in Table 4. The climate yield of maize is positively correlated with the SPEI-1 in most areas, with the SPEI-1 in July having the highest correlation with the climate yield, especially in Siping, Fuxin, Jinzhou, and Tieling, where the correlation of the Songliao Plain maize belt reaches 0.55 ($P \leq 0.01$). The relationship between the average SPEI-1 in the growing season and the climate yield is 0.72 ($P \leq 0.01$) in Fuxin city and 0.68 ($P \leq 0.01$) in the Songliao Plain maize belt. This shows that the climate yield decreases with a decrease in SPEI, that is, the drier the weather conditions are, the lower the yield is.

The Pearson correlation between the SPEI-3 in the growth season from 1998 to 2017 and the climate yield in various regions of the Songliao Plain maize belt is shown in Table 5. Except May and June, other growth season months have high correlations with yield, which indicates that the drought sensitive period in most areas is the late growth season. The SPEI-3 in August can better reflect the climate yield. The maize belt in the Songliao Plain, except for Jilin, Anshan, Benxi, Liaoyang, and Panjin, has a positive correlation with SPEI-3. There are two regions (Siping and Fuxin) with correlations of 0.72 and 0.80 ($P \leq 0.01$), and the maize belt in the Songliao Plain is 0.64 ($P \leq 0.01$). Compared with SPEI-1, SPEI-3 has a stronger correlation with the climate yield of the maize belt in the Songliao Plain.

The middle of June to the middle of August is the jointing to milking stage of maize, while the SPEI-3 in August is the cumulative result of June, July, and August in the growing season, thus the regression result in a specific month gives a better explanation than the average value in the entire growing season. Therefore, the SPEI-3 in August, the growth season with the strongest correlation, was selected for regression analysis with climate yield, and a regression equation was established:

$$y = 556.479x + 114.804 \tag{9}$$

where y is the climate yield of the maize belt in the Songliao Plain, and x is the SPEI-3 in August. Regression results confirmed that the maize climate yield was positively correlated with SPEI-3 in August during the growth season. The R^2 between the drought index and the climate yield exceeded 0.44 in 52% of Songliao Plain maize belt; $P \leq 0.01$, and the R^2 was 0.51 for the established equation.

Table 4. Pearson's correlation coefficient of SPEI-1 and climate yield in the maize growing season in the Songliao Plain maize belt during the period 1998–2017.

Month	Ha	Bai	Chang	Ji	Si	Song	An	Ben	Fush	Fux	Jin	liao	Pan	Shen	Tong	Tie	Ying	SongL
May	0.06	0.04	0.11	−0.16	0.20	0.19	−0.31	−0.26	−0.11	0.23	−0.07	−0.17	0.02	−0.06	0.38	0.14	−0.02	0.02
Jun.	0.58 *	0.31	0.31	0.19	0.33	0.43 *	0.21	0.03	0.11	0.27	0.36	0.26	0.18	0.27	0.24	0.19	0.32	0.40
Jul.	0.07	0.22	0.43 *	0.13	0.67 **	0.54 *	0.16	0.01	0.28	0.66 **	0.61 **	−0.05	0.02	0.31	−0.05	0.45	0.26	0.55 *
Aug.	0.20	0.09	0.24	0.14	0.39	0.43 *	−0.06	0.25	0.37	0.72 **	0.36	−0.06	−0.01	0.13	0.37	0.39	0.14	0.38
Sep.	0.22	0.04	0.12	−0.14	0.16	0.26	0.10	−0.13	0.02	0.47	0.38	0.08	0.05	0.22	0.38	0.16	0.25	0.27
mean	0.45 *	0.24	0.40 *	0.23	0.58 **	0.59 **	−0.01	0.09	0.03	0.72 **	0.47*	−0.24	−0.21	0.27	0.48 *	0.43 *	0.31	0.68 **

Note: * Significant at $P \leq 0.05$; ** significant at $P \leq 0.01$; red and blue indicate positive correlation and negative correlation respectively.

Table 5. Pearson correlation coefficient of SPEI-3 and climate yield in the maize growing season in the Songliao Plain maize belt during the period 1998–2017.

Month	Ha	Bai	Chang	Ji	Si	Song	An	Ben	Fush	Fux	Jin	liao	Pan	Shen	Tong	Tie	Ying	SongL
May	−0.08	0.15	−0.02	−0.18	0.17	0.14	−0.39	0.19	−0.34	0.20	−0.20	−0.20	−0.36	−0.12	0.08	0.06	−0.13	−0.14
Jun.	0.28	0.36	0.19	−0.30	0.35	0.45	−0.18	−0.35	0.29	0.30	0.12	0.26	−0.29	0.00	0.25	0.12	0.08	0.15
Jul.	0.32	0.32	0.45 *	−0.13	0.67 **	0.63 **	0.01	−0.15	0.01	0.66 **	0.47	−0.10	0.12	0.28	0.28	0.43 *	0.27	0.49
Aug.	0.39	0.30	0.50 *	0.07	0.72 **	0.66 **	0.13	0.10	0.36	0.80 **	0.64 **	0.03	−0.12	0.35	0.30	0.52 **	0.36	0.64 **
Sep.	0.18	0.15	0.38	−0.07	0.58	0.50	0.07	0.09	0.32	0.77 **	0.58 **	−0.05	0.15	0.28	0.27	0.47 *	0.26	0.51
mean	0.45	0.32	0.54 **	−0.20	0.69 **	0.57	−0.35	−0.24	−0.06	0.64 **	0.35	−0.25	−0.34	0.17	0.28	0.37	0.18	0.56

Note: * Significant at $P \leq 0.05$; ** significant at $P \leq 0.01$; red and blue indicate positive correlation and negative correlation respectively.

3.3. Analysis of Future Climate Trend and Drought Index Changes

3.3.1. Determination of the Time Range for 1.5 °C and 2.0 °C Global Warming Scenarios

The Fifth Assessment Report (IPCC.AR5) [3] pointed out that the 1.5 °C and the 2.0 °C global warming scenarios are the average temperatures from 1850 to 1900. However, owing to the limitations of the observation data, 1986–2005 was chosen as the reference period to ensure the quality of data. The average temperature in 1986–2005 was 0.61 °C higher than that before the industrial revolution (1850–1900); therefore, it needs to be further increased by 0.89 °C and 1.39 °C, which gives the global warming scenarios of 1.5 °C and 2.0 °C.

The time when global warming scenarios of 1.5 °C and 2.0 °C were achieved in this study was determined according to the results of different representative concentration pathways (RCPs) driving the global climate model. Eligible data were selected from 20 sets of data combined under five CMIP5 global climate models and four RCP scenarios. The specific calculation scheme for choosing the heating mode of 1.5 °C and 2.0 °C was as follows. First of all, the future annual global temperature simulation values were subtracted from their respective simulated mean values from 1986 to 2005, and then 0.61 °C was added to obtain the annual global warming values compared with those before industrialization, and the warming sequence was subjected to 20 year moving average processing. For the climate data of global warming of 1.5 °C, according to IPCC.AR5 and the principle of "the early warming reaches 1.5 °C and the late 21st century temperature control is within 2.0 °C", there were two eligible datasets, namely, IPSL-CM5A-LR, RCP2.6 and GFDL-ESM2M, RCP4.5. Similarly, for the climate data of global warming of 2.0 °C, two sets of data that met the conditions, namely, NorESM1-M, RCP4.5, GFDL-ESM2M, RCP6.0, were selected according to the standard of "the early warming reaches 2.0 °C and the temperature control at the end of the 21st century is within 2.5 °C". Table 6 shows the basic characteristics of the four sets of selected data.

Table 6. Basic characteristics of the selected global climate model data.

Temperature Rise	Model	Country	Horizontal Resolution	Scenarios	Time of Occurrence of Temperature Rise
1.5 °C	IPSL-CM5A-LR	France	96 × 96	RCP2.6	2021–2040 year
	GFDL-ESM2M	The United States	144 × 90	RCP4.5	2042–2061 year
2.0 °C	NorESM1-M	Norway	144 ×96	RCP4.5	2061–2080 year
	GFDL-ESM2M	The United States	144 × 90	RCP6.0	2066–2085 year

RCP: representative concentration pathways.

3.3.2. Feasibility Analysis of the Climate Model Data

Multi-model ensemble (MME) is a common method used to evaluate the simulation capability of climate models. It offsets the deviation of each model to a certain extent, is the most similar to the observation results, and is closer to the real natural state [37]. This study uses the unweighted average to calculate MME.

CMIP5 establishes the future prediction results based on the simulation of historical data so that the prediction ability is consistent with the simulation ability. In other words, a better model for simulating historical data and the future results of simulation is also credible. To understand the ability of the CMIP5 model to simulate the historical temperature and the precipitation in the maize belt of the Songliao Plain, this study used three statistics, i.e., of normalized spatial standard deviation (NSTD), Pearson's correlation (PC), and relative root mean error (RRMSE) [38], to judge the similarity between the simulated and the observed data of the model.

According to the SEO-Kyong [39] standard, the model with an NSTD value between 0.75 and 1.25 has better simulation performance. Table 7 shows that the average results of multi-model ensembles with temperatures of 1.5 °C and 2.0 °C and precipitation were better than those of a single model. The Pearson's correlations were all at the level of $P \leq 0.01$. Compared with that of temperature, the correlation of precipitation was slightly worse, but the simulation performance of the multi-mode ensemble was still higher than that of the single mode. The value of the RRMSE was negative, which indicates that the simulation level of the model was higher than the average level. The greater the negative value is, the better the simulation performance is [40]. It is not difficult to see that the RRMSE averaged by the two multi-mode sets was better than those of other single modes, and the simulation performance of the multi-mode ensemble averaged at 2.0 °C was better. Therefore, it was feasible to select the multi-mode ensemble average as the research basis for the next step.

Table 7. Normalized spatial standard deviation (NSTD), Pearson's correlation (PC), and relative root mean error (RRMSE) for observed and simulated values.

	Temperature					Precipitation				
Model	GFDL-ESM2M	IPSL-CM5A-LR	NorESM1-M	MME 1.5 °C	MME 2.0 °C	GFDL-ESM2M	IPSL-CM5A-LR	NorESM1-M	MME 1.5 °C	MME 2.0 °C
NSTD	0.874	1.13	1.39	0.99	0.93	1.48	0.91	1.26	0.98	1.02
PC	0.94 **	0.96 **	0.95 **	0.97 **	0.96 **	0.77 **	0.78 **	0.77 **	0.80 **	0.81 **
RRMSE	0.11	0.00	0.04	−0.02	−0.01	0.11	0.07	0.00	−0.03	−0.02

Note: MME1.5 °C is model IPSL-CM5A-LR and GFDL-ESM2M set average; MME2.0 °C is model NorESM1-M and GFDL-ESM2M set average. ** Significant at $P \leq 0.01$.

3.3.3. Changes in Temperature and Precipitation under the 1.5 °C and the 2.0 °C Global Warming Scenarios

The MME1.5 °C and the MME2.0 °C were used to predict future temperature and precipitation changes. Under the 1.5 °C and the 2.0 °C global warming scenarios, the change trend of temperature and precipitation in the growing season of the Songliao Plain maize belt was significant relative to the reference period (1986–2005). Under these scenarios, most areas of the Songliao Plain maize belt were mainly characterized by temperature and precipitation increase. Under the 1.5 °C scenario, the average temperature rise was 1.17 °C, and the average temperature rise was 2.14 °C under the 2.0 °C scenario; furthermore, the temperature rise amplitudes under both modes were higher than the global average. According to the geographical distribution (Figure 7a,b), the temperature rise at 2.0 °C was greater than that at 1.5 °C, especially in the northern part of the Songliao Plain maize belt. Under the two scenarios, the temperature changes in the maize belt of the Songliao Plain were less in the south and greater in the north. Under the 1.5 °C and the 2.0 °C global warming scenarios, the precipitation in the growing season of the Songliao Plain maize belt increased by 18.9 mm and 57.9 mm, respectively. From the perspective of spatial distribution (Figure 7c,d), the increase in precipitation gradually decreased from south to north, and the increase under the 2.0 °C scenario was larger than that under the 1.5 °C scenario.

3.3.4. Changes in the Drought Index under the 1.5 °C and the 2.0 °C Global Warming Scenarios

As can be seen from Figure 8, from 2020 to 2100, the SPEI fluctuated between −0.84 (2069) and 0.81 (2044) under the 1.5 °C scenario, with an average of −0.065, showing a downward trend at a rate of −0.031/10a, which indicates that the degree of drought in the maize belt of the Songliao Plain would increase. Under the 2.0 °C global warming scenario, the SPEI mean value of the Songliao Plain maize belt for the period 2020–2100 was −0.0059, with a maximum of 1.13 (2021) and a minimum of −1.11 (2065). This shows a downward trend at a rate of −0.061/10a, indicating that the drought intensity of the Songliao Plain maize belt was more serious than that under the 1.5 °C scenario.

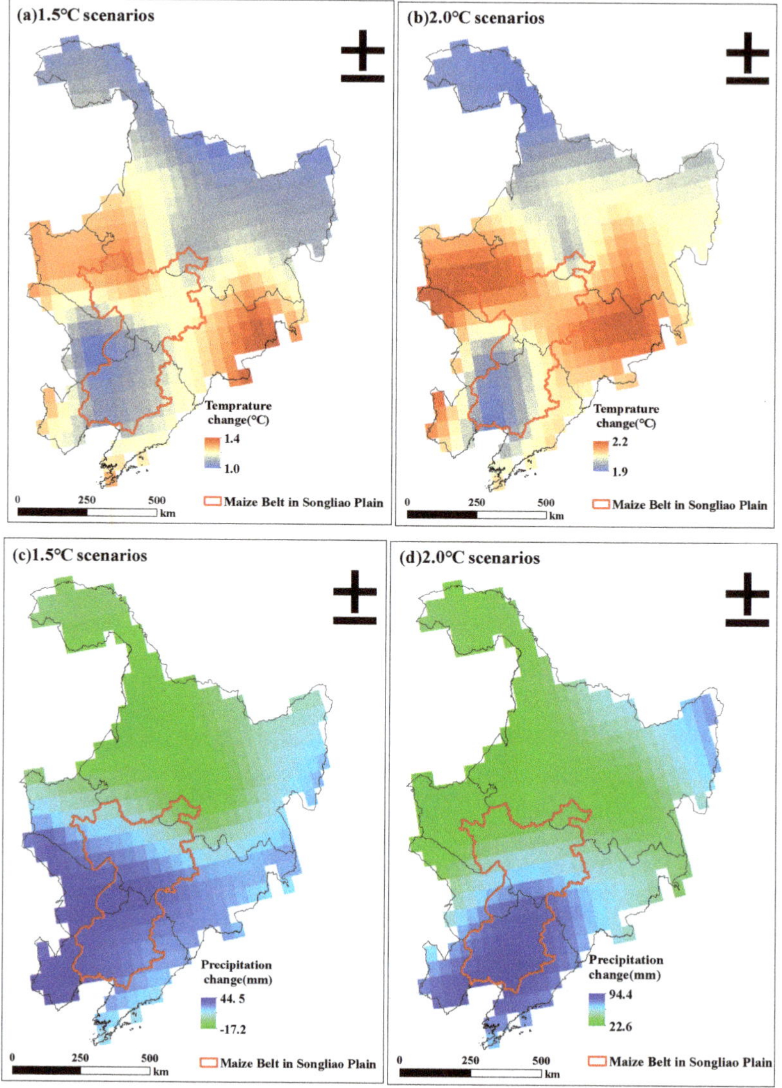

Figure 7. Distributions of temperature and precipitation in the maize growing season relative to 1986–2005 under the 1.5 °C (**a,c**) and the 2.0 °C (**b,d**) global warming scenarios.

In the future, the SPEI of the maize belt in Songliao Plain will decline. The spatial distribution of SPEI-3 is shown in Figure 9 under the 1.5 °C and the 2.0 °C global warming scenarios. When the temperature rose to 1.5 °C, SPEI-3 was between −0.28 and −0.16 compared with the reference period (1986–2005). The regions with smaller changes are located in the northern part of the Songliao Plain maize belt (Baicheng, Songyuan, Siping, Changchun, Jilin and Harbin). The change trend of SPEI-3 gradually increased from north to south, with Liaoyang and Anshan having the largest changes. SPEI-3 changed from −0.35 to −0.26 when the temperature rose to 2.0 °C. The SPEI in the eastern part of the Songliao Plain maize belt (Jilin, Changchun, Siping, Tieling, Fushun) changed little, and the greatest change occurred to the west (Baicheng, Fuxin, Jinzhou, Anshan, Yingkou, Panjin). In general, the SPEI at 2.0 °C had a large variation, about 1.5 times that of 1.5 °C. In other words, the drought in the Songliao Plain maize belt would be more serious under the 2.0 °C scenario.

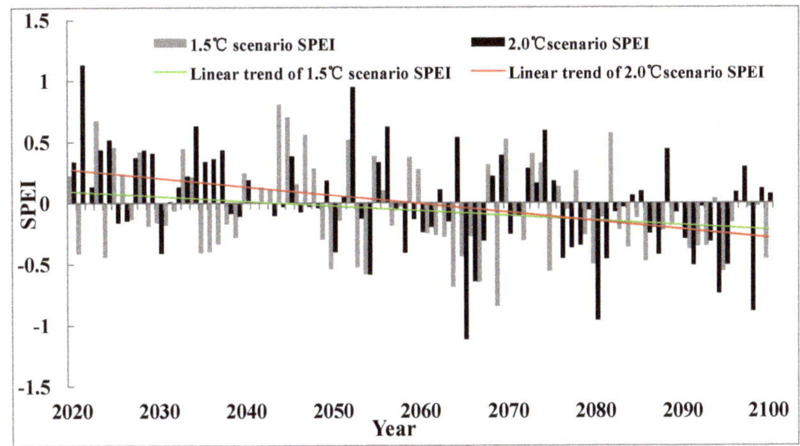

Figure 8. Temporal variation in the annual SPEI in the Songliao Plain maize belt under the 1.5 °C and the 2.0 °C global warming scenarios.

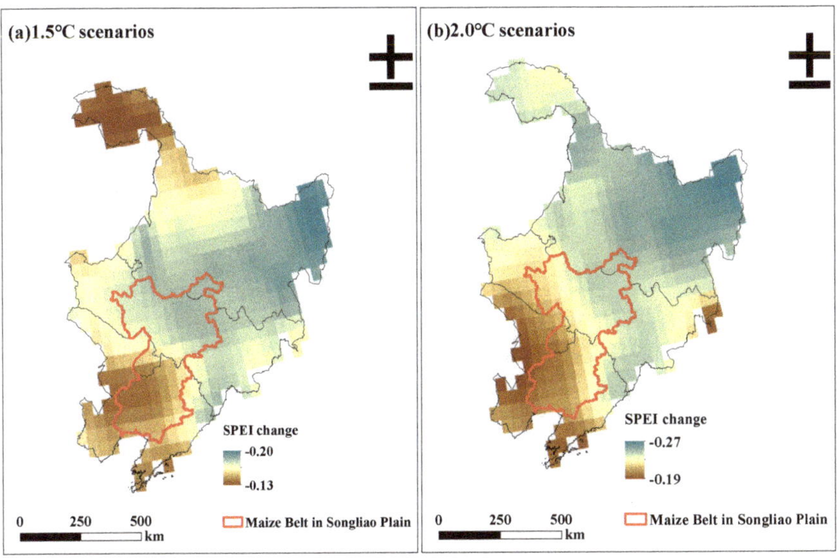

Figure 9. Spatial variation in the SPEI-3 in the Songliao Plain maize belt relative to 1986–2005 under the 1.5 °C (**a**) and the 2.0 °C (**b**) global warming scenarios.

3.4. Forecast Analysis of Future Yield

3.4.1. Climate Yield Prediction of Maize with Temperature and Precipitation Regression Under the 1.5 °C and the 2.0 °C Global Warming Scenarios

The average temperature, precipitation, and SPEI-3 accumulation in August in the growing season of the Songliao Plain maize belt simulated by CMIP5 data were substituted into the regression model obtained using historical data, and the results are shown in Figure 10. Figure 10a is the climate yield change in the Songliao Plain maize belt obtained by multiple regression of temperature and precipitation under the 1.5 °C global warming scenario with the rate of change ranging from 6.1% to −18.2%. The regions with the largest rate of change are distributed in Siping and the eastern part of Benxi, and the northwest and the northeast parts of Songliao Plain maize belt showed an increasing yield trend. Figure 10b shows the climate yield change under the 2.0 °C global warming scenario. The yield change rate ranged from 5.3% to −22.3%, with the largest yield change rate occurring in Panjin and the smallest occurring in Harbin. The yield change rate gradually increased from the northeast to the southwest of the Songliao Plain maize belt, with the northern region showing a trend of increasing production, which was more significant than the 1.5 °C scenario.

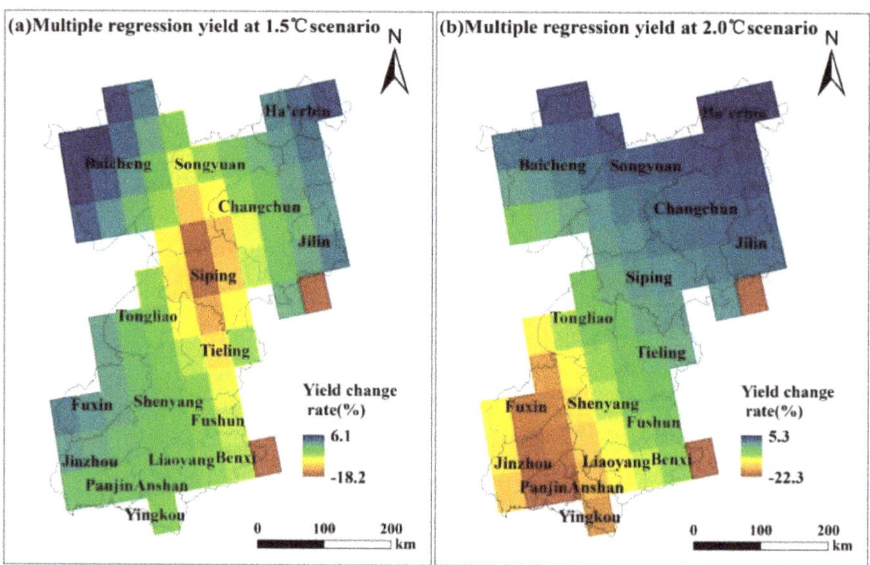

Figure 10. Distribution of the maize yield change in the Songlian Plain maize belt predicted by multiple regression under the 1.5 °C (**a**) and the 2.0 °C (**b**) global warming scenarios.

3.4.2. Climate Yield Prediction of Maize with Drought Index Regression under the 1.5 °C and 2.0 °C Global Warming Scenarios

Figure 11a is the climate yield change obtained by the August SPEI-3 one-variable regression model under the 1.5 °C global warming scenario. The climate yield change rate ranged from −5.3% to −15.1%. The regions with a high change rate were distributed in the southern part of the maize belt in Songliao Plain, and the climate yield change rate decreased further north. Figure 11b shows the climate yield change under the 2.0 °C global warming scenario. The climate yield change rate was between −17.5% and −29.2%, about twice as high as that of the 1.5 °C scenario. The yield change rate gradually increased from north to south, and this trend was similar to the spatial pattern of the 1.5 °C scenario. The overall output was still decreasing, and the yield change rate at 2.0 °C was stronger than that at 1.5 °C.

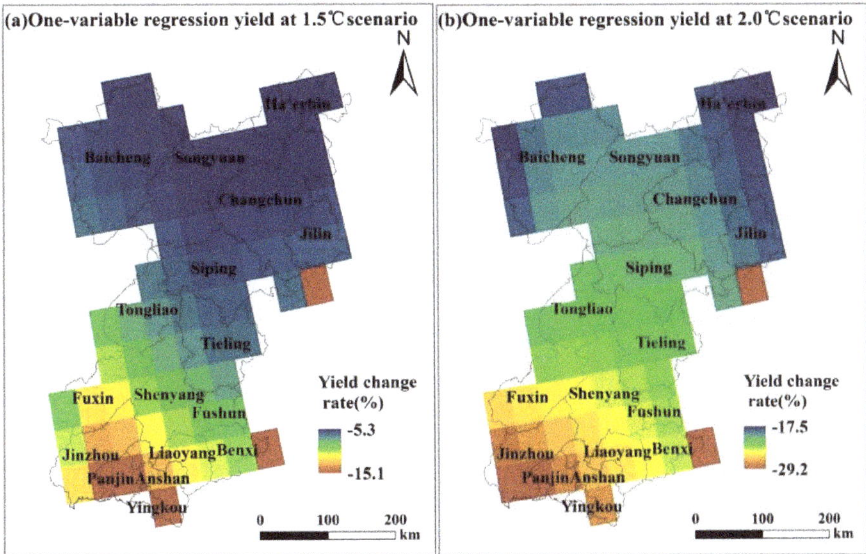

Figure 11. Distribution of the maize yield change in the Songlian Plain maize belt predicted by one-variable regression under the 1.5 °C (**a**) and the 2.0 °C (**b**) global warming scenarios.

4. Discussion

In this study, the maize belt in the Songliao Plain was taken as the research object, and the spatiotemporal changes in climate variables and the drought index during the maize growing season were studied to further establish the relationship between them and the maize climate yield. On this basis, the climate change and the yield under the scenarios of global warming of 1.5 °C and 2.0 °C were predicted. The research results could help local government departments to formulate drought and disaster reduction policies and crop spatial layout planning; moreover, the results provide a scientific basis for maize production in the Songliao Plain maize belt to adapt to future climate change. The issues that need to be further studied and discussed in the future include the following:

1. It is worth noting that, in this study, we disregarded the effect of CO_2 levels on maize yields because the year-to-year differences in CO_2 concentration are too small to generate measurable yield changes [23].

2. The effects of temperature, precipitation, and drought disaster on maize yield have been the main focuses of studies. The growth and the development of maize are very sensitive to temperature and precipitation, and the temperature requirements are different in different growth stages. Excessive or insufficient precipitation also affects the final yield of maize to different degrees. The temperature in the growing season of the Songliao Plain maize belt is negatively correlated with the maize yield. Table 2 shows that the correlation between the climate yield and the growing season temperature of the Songliao Plain maize belt in July and August is higher than that in May and June. This agrees with the findings of Bhatt et al. [41]. Waha et al. [42] found that a decrease in rainfall in the rainy season is very important for maize yield, even exceeding the effect of temperature on maize yield. This study found that the precipitation in the growing season of the maize belt in Songliao Plain has a positive effect on the maize yield, but Hawkins et al.'s [43] study found that the sensitivity of French maize yield to precipitation decreased. Labudova et al. [44] studied the standardized yields of ten crops and their correlations with SPEI and SPI on a one-month, two-month and three-month scale; the results revealed that the highest correlation can be seen between maize yield and three-month scale SPEI in August. This finding is consistent with

the results of this study. With the continuous development of the economy and technology, the influence of human social activities is increasing. In addition to temperature, precipitation, and dry and wet conditions, there are also different drip irrigation modes [45], mulching conditions [46], planting densities [47], balanced fertilizers such as nitrogen and phosphorus [48], adaptabilities of maize itself to the environment, and crop management methods, which all affect the yield of maize. Therefore, a multi-factor evaluation model that can be used to evaluate climate variables and their impact on yield remains to be constructed.

3. There are many studies on climate change trend prediction and threshold under the background of global warming scenarios of 1.5 °C and 2.0 °C, but those on its impact on the yield and the physiological processes of major grain crops are still relatively few. In particular, the risks faced by different grain producing areas in China under the background of temperature rises of 1.5 °C and 2.0 °C are still in a state of continuous research. As the global academic community has not reached a consensus on how to define the global warming of 1.5 °C and 2.0 °C, the current climate prediction and impact research on the global warming of 1.5 °C and 2.0 °C mostly adopts the multi-mode ensemble average method [49–51] to obtain the warming response under the instantaneous change condition rather than the temperature rise under the stable state expected over the long term. Researchers still need to design a model prediction test specifically for the global warming scenarios of 1.5 °C and 2.0 °C to form a special scenario and provide support for impact prediction in different fields.

4. The climate simulation model has uncertainty in setting parameters and estimating greenhouse gas emission physical processes. There are also uncertainties in the social development model and government intervention in greenhouse gas emissions. The combination of these uncertainties magnify the uncertainties that climate change may bring. In this study, five climate models were selected from many CMIP5 models according to the ISI-MIP recommendation. Global warming reached 1.5 °C and 2.0 °C during 1986–2005, which was used as the reference period. Compared with other climate models, the selected model can more effectively support impact assessment in different fields and generate more credible results. Considering the four RCP scenarios and the conditions of stable temperature rises of 1.5 °C and 2.0 °C at the end of the 21st century, the time periods and THE data of temperature rises of 1.5 °C and 2.0 °C under different model scenarios were selected to minimize the uncertainty of the climate data. This study found that the risk of maize yield reduction in the Songliao Plain maize belt under the background of global warming of 2.0 °C was significantly higher than that under the 1.5 °C scenario. Summing up the previous global warming research results, the greater the global warming amplitude is, the greater the comprehensive harmful impact of global warming on maize production will be, which may be due to the large evapotranspiration caused by the increase in the temperature rise amplitude. Although the total precipitation is increasing, the evapotranspiration caused by the warming is more severe. In other words, the excessive warming amplitude offsets the positive impact of the precipitation; therefore, frequent droughts during the maize growing season may affect the maize yield, which may also be the possibility of extreme weather events such as high temperature disasters and heavy rainfall with the 1.5 °C global warming and the 2.0 °C global warming scenarios becoming more and more obvious, leading to an increased risk of maize yield reduction.

5. Conclusions

In this paper, a relationship between historical meteorological data and yield data was determined. CMIP5 data were used to study the variation characteristics of temperature, precipitation, and drought in the future growth season and then to predict the change in maize climate yield in the Songliao Plain maize belt under the 1.5 °C and the 2.0 °C global warming scenarios.

1. The temperature in the growing season of the maize belt in the Songliao Plain had an increasing trend and the precipitation had a decreasing trend from 1965 to 2017. Geographically, Harbin, Liaoyang, and Anshan had the highest temperature rise, while the southeast of the Songliao

Plain maize belt had the largest precipitation reduction. SPEI-1 and SPEI-3 in the maize belt of the Songliao Plain were very sensitive to temperature and precipitation and were calculated by using the September SPEI-6. The frequencies of light drought, moderate drought, severe drought, and extreme drought were 16.0%, 10.4%, 5.8%, and 2.0%, respectively.

2. From 1998 to 2017, 65% of the maize belt in the Songliao Plain showed a trend of yield reduction, with the lowest yield occurring in 2000. The unit yield in the northeast was higher than that in the southwest Songliao Plain maize belt in spatial distribution. From 1998 to 2017, the climate yield was negatively correlated with the temperature in the growing season and positively correlated with the precipitation; furthermore, it was positively correlated with the drought index SPEI-3, and the correlation with SPEI-3 was higher.

3. Under the 1.5 °C and the 2.0 °C global warming scenarios, the temperature and the precipitation in the growing season of the maize belt in the Songliao Plain exhibited upward trends relative to the reference period (1986–2005). In the spatial distribution, the temperature increased from south to north, while the precipitation decreased from south to north. These results reveal that, as a whole, drought in the Songliao Plain Maize Belt will be more serious in the future.

4. Under the 1.5 °C and the 2.0 °C global warming scenarios, the climate yield changes in the Songliao Plain maize belt predicted by meteorological factor regression were −7.7% and −15.9%, respectively. The climate yield changes obtained by drought index regression were −12.2% and −21.8%, respectively. The more serious the drought is, the more negative its effects on maize yield are.

Author Contributions: All authors contributed meaningfully to this study. A.G. and J.Z. conceived the research topic. S.T., A.H., and Y.B. were responsible for designing the methodology, data acquisition, and analysis. K.L. provided methodology support, and A.G. drafted the manuscript. All authors read and revised the final manuscript.

Funding: This research is supported by the National Science Foundation of China (41571491 and 41877520); the Science and Technology Development Planning of Jilin Province (20190303018SF); and the Science and Technology Planning of Changchun (19SS007).

Acknowledgments: We are most grateful to the National Meteorological Information Center and Compled Model Inter-comparision Project (CMIP5) for kindly providing the climate data and information. The authors are grateful to the many individuals working on the development of free and open-source software for supporting the sharing of knowledge.

Conflicts of Interest: The authors declare no conflicts of interest.

References

1. Miah, M.G.; Abdullah, H.M.; Jeong, C. Exploring standardized precipitation evapotranspiration index for drought assessment in Bangladesh. *Environ. Monit. Assess.* **2017**, *189*, 547. [CrossRef] [PubMed]

2. Solomon, S.; Qin, D.; Manning, M.; Chen, Z.; Marquis, M.; Averyt, K.B.; Tignor, M.; Miller, H.L. *Climate Change 2007: The Physical Science Basis. Contribution of Working Group I to the Fourth Assessment Report of the Intergovernmental Panel on Climate Change*; Cambridge University Press: Cambridge, NY, USA, 2007.

3. Stocker, T.F.; Dahe, Q.; Plattner, G.K. *Climate Change 2013: The Physical Science Basis. Working Group I Contribution to the Fifth Assessment Report of the Intergovernmental Panel on Climate Change*; Summary for Policymakers (IPCC, 2013); IPCC: Bern, Switzerland, 2013; pp. 1–33.

4. Lambi, C.; Forbang, T. The economic impact of climate change on agriculture in cameroon. *Earth Environ. Sci.* **2009**, *6*, 092017. [CrossRef]

5. Guo, J.P. Advance in impacts of climate change on agricultural production in China. *J. Appl. Meteor Sci.* **2015**, *26*, 1–11.

6. Adams, R.M.; Hurd, B.H.; Lenhart, S.; Leary, N. Effects of global climate change on agriculture: On interpretative review. *Clim. Res.* **1998**, *11*, 19–30. [CrossRef]

7. Babel, M.S.; Agarwal, A.; Swain, D.K.; Herath, S. Evaluation of climate change impacts and adaptation measures for rice cultivation in Northeast Thailan. *Clim. Res.* **2011**, *46*, 137–146. [CrossRef]

8.   Challinor, A.; Watson, J.; Lobell, D.; Howden, S.; Smith, D.; Chhetri, N. A meta-analysis of crop yield under climate change and adaptation. *Nat. Clim. Chang.* **2014**, *4*, 287–291. [CrossRef]

9.   Sisodia, B.V.S.; Yadav, R.R.; Kumar, S.; Sharma, M.K. Forecasting of pre-harvest crop yield using discriminant function analysis of meteorological parameters. *J. Agrometeorol.* **2014**, *16*, 121–125.

10.   Brisson, N.; Gate, P.; Gouache, D. Why are wheat yields stagnating in Europe? A comprehensive data analysis for France. *Field Crops Res.* **2010**, *119*, 201–212. [CrossRef]

11.   Wei, T.; Cherry, T.L.; Glomrod, S.; Zhang, T. Climate change impacts on crop yield: Evidence from China. *Sci. Total Environ.* **2014**, *499*, 133–140. [CrossRef]

12.   Lobell, D.B.; Schlenker, W.; Costa-Roberts, J. Climate trends and global crop production since 1980. *Science* **2011**, *333*, 616–620. [CrossRef]

13.   Gupta, R.; Mishra, A. Climate change induced impact and uncertainty of rice yield of agro-ecological zones of India. *Agric. Prod. Syst.* **2019**, *173*, 1–11. [CrossRef]

14.   Chen, C.; Pang, Y.M.; Pan, X.B.; Zhang, L. Impacts of climate change on cotton yield in china from 1961 to 2010 based on provincial data. *Environ. Monit. Assess.* **2014**, *29*, 515–552. [CrossRef]

15.   Huang, J.P.; Yu, H.P.; Guan, X.D.; Wang, G.; Guo, R. Accelerated dry land expansion under climate change. *Nat. Clim. Chang.* **2016**, *6*, 166. [CrossRef]

16.   Peng, S.B.; Huang, J.L.; Sheehy, J.E.; Laza, R.C.; Visperas, R.M.; Zhong, X.; Centeno, G.S.; Khush, G.S.; Cassman, K.G. Rice yields decline with higher night temperature from global warming. *Proc. Natl. Acad. Sci. USA* **2004**, *101*, 9971–9975. [CrossRef] [PubMed]

17.   Lobell, D.B.; Field, C.B. Global scale climate–crop yield relationships and the impacts of recent warming. *Environ. Res. Lett.* **2007**, *2*, 014002. [CrossRef]

18.   Chen, Y.; Zhang, Z.; Tao, F. Impacts of climate change and climate extremes on major crops productivity on China at a global warming of 1.5 °C & 2.0 °C. *Earth Syst. Dynam.* **2018**, *9*, 543–562.

19.   Van Rooy, M.P. A rainfall anomaly index independent of time and space. *Notos* **1965**, *14*, 43–48.

20.   Palmer, W.C. *Meteorological Drought: US Department of Commerce*, 3rd ed.; Weather Bureau: Washington, DC, USA, 1965.

21.   McKee, T.B.; Doesken, N.J.; Kleist, J.; Anaheim, C.A. The relationship of drought frequency and duration to time scales. In Proceedings of the 8th Conference on Applied Climatology, Boston, MA, USA, 17–22 January 1993.

22.   Zhang, Q.; Zhou, X.K.; Xiao, F.J. *GB/T20481-2006, Meteorological Drought Level, National Standards of the People's Republic of China*; China Standars Publisher: Beijing, China, 2006.

23.   Vicente-Serrano, S.M.; Begueria, S.; Lopezmoreno, J.I. A multi-scalar drought index sensitive to global warming: The standardized precipitation evapotranspiration index. *J. Clim.* **2010**, *23*, 1696–1718. [CrossRef]

24.   Zhang, Q.; Hu, Z. Assessment of drought during maize growing season in Northeast China. *Theor. Appl. Climatol.* **2018**, *133*, 1315–1321. [CrossRef]

25.   Matiu, M.; Ankerst, D.P.; Menzel, A. Interactions between temperature and drought in global and regional crop yield variability during 1961–2014. *PLoS ONE* **2017**, *12*. [CrossRef]

26.   Zhang, J.Q. Risk assessment of drought disaster in the maize-growing region of songliao Plain, China. *Agric. Ecosyst. Environ.* **2004**, *102*, 133–153. [CrossRef]

27.   Vicente-Serrano, S.M.; Cabello, D.; Tomas-Burguera, M.; Martín-Hernández, N.; Beguería, S.; Azorin-Molina, C.; Kenawy, A. Drought variability and land degradation in semiarid regions: Assessment using remote sensing data and drought indices (1982–2011). *Remote Sens.* **2015**, *7*, 4391–4423. [CrossRef]

28.   Zhang, Q.; Zhang, J.Q. Drought hazard assessment in typical maize cultivated areas of china at present and potential climate change. *Nat. Hazards* **2016**, *81*, 1323–1331. [CrossRef]

29.   Zhang, Q.; Zhang, J.Q.; Wang, C. Risk assessment of drought disaster in typical area of maize cultivation in China. *Theor. Appl. Climatol.* **2017**, *128*, 533–540. [CrossRef]

30.   Mann, H.B. Nonparametric tests against trend. *Econometrica* **1945**, *13*, 245–259. [CrossRef]

31.   Kendall, M.G. *Rank Correlation Methods*; Hafner: New York, NY, USA, 1962.

32.   Xing, L.; Huang, L.; Chi, G.; Yang, L.; Li, C.; Hou, X. A dynamic study of a karst spring based on wavelet analysis and the mann-kendall trend test. *Water* **2018**, *10*, 698. [CrossRef]

33.   Lobell, D.B.; Asner, G.P. Climate and management contributions to recent trends in U.S. Agricultural Yields. *Science* **2003**, *299*, 1032. [CrossRef]

34. Guo, E.L.; Liu, X.P.; Zhang, J.Q.; Wang, Y.; Wang, C.; Wang, R.; Li, D. Assessing spatiotemporal variation of drought and its impact on maize yield in Northeast China. *J. Hydrol.* **2017**, *553*, 231–247. [CrossRef]

35. Mitra, S.; Srivastava, P. Spatiotemporal variability of meteorological droughts in southeastern USA. *Nat. Hazards* **2017**, *86*, 1007–1038. [CrossRef]

36. Prabnakorn, S.; Maskey, S.; Suryadi, F.X.; de Fraiture, C. Rice yield in response to climate trends and drought index in the Mun River Basin, Thailand. *Sci. Total Environ.* **2018**, *621*, 108–119. [CrossRef]

37. Jiang, L.; Yao, Z.; Huang, H.Q. Climate variability and change on the Mongolian plateau: Historical variation and future predictions. *Clim. Res.* **2016**, *67*, 1–14. [CrossRef]

38. Sillmann, J.; Kharin, V.V.; Zhang, X.; Bronaugh, D. Climate extremes indices in the CMIP5 multi model ensemble: Part1. Model evaluation in the present climate. *J. Geophys. Res.* **2013**, *118*, 1716–1733.

39. SEO, K.H.; Jung, O. Assessing future change in the East Asian summer monsoon using CMIP3 Models: Results from the best model ensemble. *J. Clim.* **2013**, *26*, 1807–1817. [CrossRef]

40. Gleckler, P.J.; Taylor, K.E.; Doutriaux, C. Performance metrics for climate models. *J. Geophys. Res.* **2008**, *113*, 1–20. [CrossRef]

41. Bhatt, D.; Maskey, S.; Babel, M.S.; Uhlenbrook, S.; Prasad, K.C. Climate trends and impacts on crop production in the Koshi River Basin of Nepal. *Reg. Environ. Chang.* **2014**, *14*, 1291–1301. [CrossRef]

42. Waha, K.; Muller, C.; Rolinki, S. Separate and combined effects of temperature and precipitation change on maize yields in sub-saharan Africa for mid-to late-21st century. *Glob. Planet. Chang.* **2013**, *106*, 1–12. [CrossRef]

43. Hawkins, E.; Fricker, T.E.; Challinor, A.J.; Ferro, C.A.; Ho, C.K.; Osborne, T.M. Increasing influence of heat stress on French maize yields from the 1960s to 2030s. *Glob. Chang. Biol.* **2013**, *19*, 937–947. [CrossRef]

44. Labudova, L.; Labuda, M.; Takac, J. Comparison of SPI and SPEI applicability for drought impact assessment on crop production in the danubian lowland and the east Slovakian Lowland. *Theor. Appl. Climatol.* **2017**, *128*, 491–506. [CrossRef]

45. Zhou, L.; Feng, H.; Zhao, Y.; Qi, Z.; Zhang, T.; He, J.; Dyck, M. Drip irrigation lateral spacing and mulching effects the wetting pattern, shoot-root regulation, and yield of maize in a sand-layered soil. *Agric. Water Manag.* **2017**, *184*, 114–123. [CrossRef]

46. Qin, X.L.; Li, Y.Z.; Han, Y.L.; Hu, Y.; Li, Y.; Wen, X.; Liao, Y.; Siddique, K.H. Ridge-Furrow mulching with black plastic film improves maize yield more than white plastic film in dry areas with adequate accumulated temperature. *Agric. For. Meteorol.* **2018**, *262*, 206–214. [CrossRef]

47. Wetzel, J.; Stone, A. Yield response of winter squash to irrigation regime and planting density. *HortScience* **2019**, *54*, 1190–1198. [CrossRef]

48. Zhang, Y.; Wang, R.; Wang, H.; Wang, S.; Wang, X.; Li, J. Soil water use and crop yield increase under different long-term fertilization practices incorporated with two-year tillage rotations. *Agric. Water Manag.* **2019**, *221*, 362–370. [CrossRef]

49. Alfieri, L.; Bisselink, B.; Dottori, F.; Naumann, G.; de Roo, A.; Salamon, P.; Wyser, K.; Feyen, L. Global projections of river flood risk in a warmer world. *Earth's Future* **2016**, *5*, 171–182. [CrossRef]

50. Taylor, K.E.; Stouffe, B.J.; Meehl, G.A. An overview of CMIP5 and the experiment design. *Bull. Am. Meteorol. Soc.* **2012**, *93*, 485–498. [CrossRef]

51. Zhou, B.T.; Wen, H.Q.Z.; Xu, Y.; Song, L.; Zhang, X. Projected changes in temperature and precipitation extremes in China by the CMIP5 multimodel ensembles. *J. Clim.* **2014**, *27*, 6591–6611. [CrossRef]

Article

Seasonal Drought Pattern Changes Due to Climate Variability: Case Study in Afghanistan

Ishanch Qutbudin [1], Mohammed Sanusi Shiru [1,2], Ahmad Sharafati [3], Kamal Ahmed [1,4],
Nadhir Al-Ansari [5], Zaher Mundher Yaseen [6,*], Shamsuddin Shahid [1] and Xiaojun Wang [7,8]

[1] Department of Water and Environmental Engineering, School of Civil Engineering, Faculty of Engineering, Universiti Teknologi Malaysia (UTM), 81310 Johor Bahru, Malaysia; 8884ishanch@gmail.com (I.Q.); shiru.sanusi@gmail.com (M.S.S.); kamal_brc@hotmail.com (K.A.); sshahid@utm.my (S.S.)

[2] Department of Environmental Sciences, Faculty of Science, Federal University Dutse, P.M.B 7156 Dutse, Nigeria

[3] Department of Civil Engineering, Science and Research Branch, Islamic Azad University, Tehran, Iran; asharafati@srbiau.ac.ir

[4] Faculty of Water Resource Management, Lasbela University of Agriculture, Water and Marine Sciences, Balochistan 90150, Pakistan

[5] Civil, Environmental and Natural Resources Engineering, Lulea University of Technology, 97187 Lulea, Sweden; nadhir.alansari@ltu.se

[6] Sustainable Developments in Civil Engineering Research Group, Faculty of Civil Engineering, Ton Duc Thang University, Ho Chi Minh City, Vietnam

[7] State Key Laboratory of Hydrology-Water Resources and Hydraulic Engineering, Nanjing Hydraulic Research Institute, Nanjing 210029, China; xjwang@nhri.cn

[8] Research Center for Climate Change, Ministry of Water Resources, Nanjing 210029, China

* Correspondence: yaseen@tdtu.edu.vn; Tel.: +84-033-498-7030

Received: 20 April 2019; Accepted: 23 May 2019; Published: 25 May 2019

Abstract: We assessed the changes in meteorological drought severity and drought return periods during cropping seasons in Afghanistan for the period of 1901 to 2010. The droughts in the country were analyzed using the standardized precipitation evapotranspiration index (SPEI). Global Precipitation Climatology Center rainfall and Climate Research Unit temperature data both at 0.5° resolutions were used for this purpose. Seasonal drought return periods were estimated using the values of the SPEI fitted with the best distribution function. Trends in climatic variables and SPEI were assessed using modified Mann–Kendal trend test, which has the ability to remove the influence of long-term persistence on trend significance. The study revealed increases in drought severity and frequency in Afghanistan over the study period. Temperature, which increased up to 0.14 °C/decade, was the major factor influencing the decreasing trend in the SPEI values in the northwest and southwest of the country during rice- and corn-growing seasons, whereas increasing temperature and decreasing rainfall were the cause of a decrease in SPEI during wheat-growing season. We concluded that temperature plays a more significant role in decreasing the SPEI values and, therefore, more severe droughts in the future are expected due to global warming.

Keywords: meteorological drought; standardized precipitation evapotranspiration index; climate variability; seasonal drought; drought return period

---

1. Introduction

The changes in global energy balance due to warming have changed the patterns of the atmospheric variables [1–3]. The global temperature increases, in particular, have influenced the occurrence of droughts as well as their frequency and severity in many regions [4–6]. Water availability is the main element that changes in response to changes in drought patterns [7]; thus, it is an important factor for defining water stress, agricultural productivity, and food security [8–11]. Compared to other natural disasters, droughts are unique due to their slow onset and their often prolonged occurrence [12]. As its effects accumulate slowly over time, the determination of drought onset, duration, and termination is ambiguous [13]. Therefore, the impacts of droughts on agriculture are much more devastating compared to other natural disasters [14]. The 1998 drought in Oklahoma, USA caused loses in agriculture exceeding USD $2 billion [15]. Droughts caused severe famine in the horn of Africa between 2011 and 2012, and projections have shown the likelihood of increased droughts in future [6], which would affect agriculture [16].

The variability in climate has devastating impacts on droughts characteristics, particularly in areas that are arid or semi-arid. Hence, the consequences of climate change are remarkably affecting drought patterns, having far-reaching impacts in social, economic, agricultural, and environmental contexts [17,18]. Comprehending the association between climate variability and drought pattern is vital for discerning the changes in droughts due to changes in climate [13]. However, this relationship highly varies from one region to another [19], which emphasizes the need for more region-specific studies. In addition, due to crop water demand during cropping seasons, occurrence of droughts during these seasons can be more devastating. Therefore, understanding meteorological droughts during different cropping seasons is paramount for sustainable agricultural practices.

Afghanistan, being a semi-arid to arid area, is prone to droughts and prolonged droughts commonly occur in the country. The country receives a meagre annual precipitation of between 200 and 400 mm [20], indicating the significance of natural resource management and the requirement to adapt and implement mitigation measures against climate change. The changing climate in the country significantly affects most rural communities where water, soil, forests, and grazing areas are essential factors [21]. The impacts of climate change on water resources have been reported in Afghanistan; observation showed the impact of climate change on glacier and snowmelts feeding the Kabul River, which has caused an increase in the trend of melting and a shift in the seasonal monsoons in the river basin [21].

Studies related to droughts in Afghanistan are limited. Muhammad et al. [22] applied a modified version of standard Palmer method, incorporating snowmelt module based on the temperature-index method to assess droughts between 1957 and 2002. The study revealed major drought events over the region during the period. Droughts were found to have severe impacts on crop yields, causing economic crisis in the province of Badakhshan of Afghanistan [23]. Williams [24] reported the increasing impacts of the changing climate in Afghanistan, as seen in the recurrent droughts, which have increased in duration and frequency, at least for the past second half of the 20th century. The severity of droughts in the neighboring Central Asia and other countries bordering Afghanistan have been reported to have increased [21,25,26]. In Central Asia, Ta [27] analyzed the spatio-temporal patterns of conditions of dryness and wetness using the standardized precipitation evapotranspiration index (SPEI) and reported a drought frequency of 42.87% in the region during 1930 to 2014. Li et al. [28] reported a sharp temperature increase was experienced in Central Asia in 1997, and has since been in a state of high volatility, making the last 15 years (1998–2013) the warmest on record in about 50 years. Zoljoodi and Didevarasl [29] assessed the trends in droughts in the northwest and northeast parts of Iran bordering Afghanistan during 1951–2005 using the Palmer Drought Severity Index, and reported an increase in the severity of droughts. Ahmed et al. [26] investigated the trends in droughts in Pakistan over the last century and reported that drought-affected areas in the country have increased due to sharp temperature increases. However, no study has assessed the meteorological droughts in Afghanistan during cropping seasons, and no attempt has been made to understand the changes in droughts in the country due to climate change.

The continuous increase in drought frequency may have devastating effects on the water supplies in many communities in the country, thus leading to humanitarian crises, including diseases, population displacements, and violent conflicts [30]. Droughts may also increase the tension between Afghanistan and neighboring countries as the various countries would try to claim a greater share of the region's total available water [31]. For Afghanistan, where a significant part of the population is engaged in agriculture, assessment of the impacts of climate change on drought characteristics is paramount for sustainable agricultural practices. This is especially crucial for the different crop growing seasons due to the changing patterns of the climate, which, for example, drastically prematurely reduce winter snowpack from its usual conditions, affecting wheat production yields in the northern, northeastern, and western regions of the country from 2007 to 2008 [30].

In the present study, the changing drought characteristics during different cropping seasons (rice, corn, and wheat) in Afghanistan during 1901–2010 were assessed. We assessed the droughts' characteristics using the SPEI [32], which considers temperature and precipitation climatic variables. Gridded precipitation and temperature datasets from the Global Precipitation Climatology Centre (GPCC) [33] and Climatic Research Unit (CRU) [34], respectively, were used for drought assessment during 1901–2010 using a 50-year sliding window with a 10-year interval. The modified Mann–Kendall (MMK) trend test was used to perform model analysis on the secular climatic variables' trends and the drought index.

2. Study Area and Data

2.1. Description of the Study Area

Afghanistan (29–39° N; 60–75° E) is a landlocked country, sharing a long border with Pakistan in the south; Iran in the west; Turkmenistan, Uzbekistan, and Tajikistan in the north; and a thin stretch mountainous land with China in the northeast (Figure 1). The Hindu Kush Mountains from the northeastern border to west across the country divide the topography of the country into geographic regions: central highland, the southwestern plateau, and the northern plains.

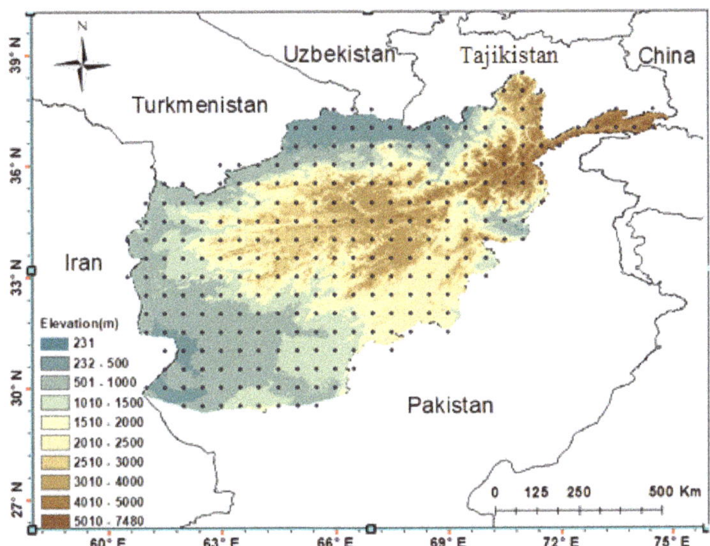

Figure 1. Map of Afghanistan showing elevation and country borders.

Afghanistan's climate varies greatly between topographic regions [35]. The climate varies considerably between arid and semiarid with hot summers and cold winters. Mean annual precipitation for the country is shown in Figure 2a, which indicates a gradual decrease in rainfall from the northeast along the Hindokush Mountains toward the southwestern plateau zone. The lowest mean annual precipitation of around 30 mm occurs in the southwestern plateau region. The highest precipitation of more than 1000 mm occurs in the northeastern Hindokush Mountain's foothill regions. The mean annual maximum temperature in Afghanistan is shown in Figure 2b. Temperature gradually increases from the top of the Hindokush mountain from northeast toward the southeastern plateau regions for both cropping seasons of winter and summer. The highest mean annual maximum temperatures are recorded in the lower regions in the southeastern zone.

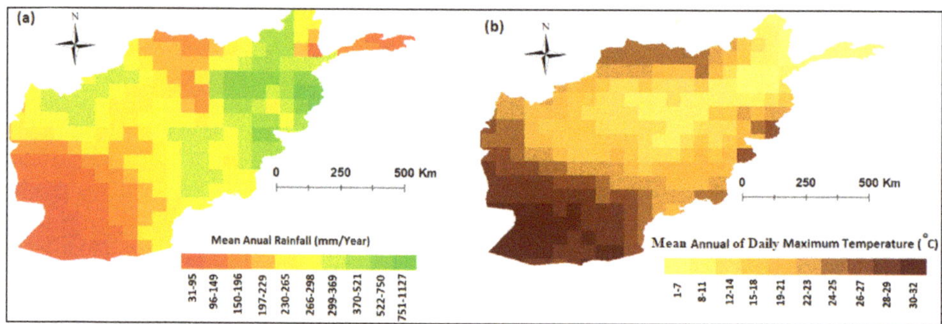

Figure 2. (**a**) Geographical distribution of mean annual rainfall (mm/year) and (**b**) annual mean of daily maximum temperature (°C) in Afghanistan.

Afghanistan is largely dependent on agricultural production. Agricultural crops depend on specific climate conditions, and changes to the climate can have crucial negative impacts on the ways in which crops are cultivated in rain-fed areas. Almost 67–85% of the Afghan population are engaged in farming [36]. Crops are cultivated during the summer and winter season. The summer cropping season starts in May and finishes in November. The main crops grown during summer are rice and corn. The main cultivated crops in winter (October to June) are wheat and barley. Wheat, rice, and corn, analyzed in this study, are the most important food crops for the population. The area cultivated for wheat is 2,575,000 ha or 79% of the total cultivated land; 200,000 ha rice, or 6.2%; and 140,000 ha corn, or 4.3% of the cultivated land in Afghanistan.

The cropping season calendars of the most important crops cultivated in the country, wheat, corn, and rice, are presented in Figure 3. Wheat is cultivated in winter, whereas rice and corn are cultivated in summer. Irrigated wheat sowing usually begins late October and finishes in early December. Rain-fed cropping season starts in winter when enough soil moisture content has been acquired from rainfall. Due to cold winters, winter season crop harvesting starts in late May and ends early July. Summer season starts in April and continues until the end of winter season in early July. In Afghanistan, wheat is cultivated in every part of the country, whereas rice and corn are cultivated in the northern part of the country. Due to adequate surface water in the north of Afghanistan, it is used as the main source of irrigation for summer crops.

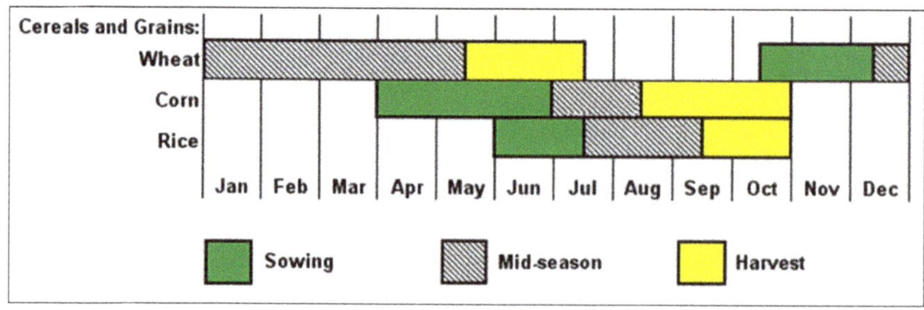

Figure 3. Crop calendars of selected crops in Afghanistan.

2.2. Datasets

One main challenge in hydro-climatological studies around the world is inadequate or lacking long-term records and reliable climatic data [26]. Prolonged conflicts that have resulted in inadequate data availability in Afghanistan have hindered climatic research in the country [37]. In areas like Afghanistan where long-tern climate data are lacking, gridded climate data are used for climatic studies [38]. Amongst gridded data, such as satellite based gridded data, the in-situ-based gridded data, reanalysis, and combined gridded data, in-situ-based gridded data have been recommended because of better spatial and temporal availability [39].

Gridded monthly precipitation data of the GPCC and the monthly average temperature data of the CRU, both at a spatial resolutions of 0.5° by 0.5° (281 grid points over Afghanistan) for the period 1901–2010, were used in this study. Data spanning the period 1901–2010 were chosen due to the use of a 50-year sliding window with a 10-year interval, which enabled uniform division and consistency within the considered period. GPCC monthly data were developed from various data sources, including national weather services and hydrological institutes with data collections from over 8500 rain gauge stations in 190 countries around the world [40]. GPCC gridded data have many advantages: (1) good dataset quality; (2) it is a dataset of a climate model, interpolated with the highest number of observed data; and (3) it has a large enough time span for hydro-climatic investigations [41]. The CRU dataset was derived from gauge measurements of 4000 weather stations from around the globe [26]. The CRU monthly temperature dataset is updated and controlled by manual and semi-automatic quality control means [34].

3. Methodology

The procedure adopted to estimate the changes in droughts and assess the influence of climatic factors on the trends in droughts during Afghanistan's major crop growing seasons is outlined below:

1. The potential evapotranspiration (PET) was estimated using the Thornthwaite method at each CRU grid point. CRU's monthly average mean temperature was used for this purpose.
2. The SPEI was calculated using GPCC rainfall and the ET estimated in Step 1 at every GPCC/CRU grid point in Afghanistan (281 grid points) during 1901–2010 using a moving 50-year window and a 10-year interval.
3. Trends in temperature, precipitation, and SPEI were assessed for cropping seasons for different 50-year periods using the MMK test.
4. The return periods (RPs) of the different drought intensities (moderate, severe, and extreme) for cropping seasons for different 50-year periods were calculated through fitting the SPEI values with an appropriate probability distribution function (PDF).
5. The results obtained for the different crops growing seasons during 1901–2010 were analyzed to understand the impacts of climate change on droughts.

The methods employed in the current study are detailed in subsequent sections.

3.1. Standardized Precipitation Evapotranspiration Index for Estimating Seasonal Droughts

PET was estimated using different methods based on temperature, radiation, and mass transfer [42]. According to Stagge et al. [43], estimated SPEI does not vary significantly with the PET method used. However, Begueria et al. [44] reported significant variations in SPEI calculated using different PET methods in arid and semi-arid regions. Begueria et al. [44] proposed Penman–Monteith as the best option for the estimation of PET for calculation of SPEI followed by Hargreaves and the Thornthwaite methods. However, the Penman–Monteith method requires sunshine, humidity, and wind speed data along with temperature data, which are all spatially and temporally inadequate in Afghanistan. The Thornthwaite method only requires the mean daily temperature and latitudes of the study area.

This makes Thornthwaite a suitable method for calculation of PET in data-scarce regions such as Afghanistan. Therefore, the Thornthwaite method was used in this study to calculate PET:

$$\text{PET} = 16 \text{N}_\text{m} \left(\frac{10 \text{T}_\text{m}}{I} \right)^{a}$$

(1)

where N_m is a correction factor estimated according to latitude of the location and the month of the year; T_m is the monthly average of daily mean temperature (°C); I is a heat index, which was estimated for the whole year; and a is a coefficient estimated from I.

For calculation of SPEI, precipitation and PET data were used for the preparation of the time series of water deficit or surplus (D) as precipitation minus potential evapotranspiration (P–PET) P–PET at each GPCC/CRU grid point. The D values were then fit with the best PDF. The D can have negative values and, therefore, three-parameter distributions are needed to calculate the SPEI. The log-logistic distribution was found to fit the D values best for all the cropping seasons. Therefore, the probability distribution function of D was estimated using a log-logistic distribution:

$$\text{F}(D) = \left[1 + \left(\frac{\alpha}{x - \gamma} \right)^{\beta} \right]^{-1}$$

(2)

where α, β, and γ are scale, shape, and origin parameters, respectively. The SPEI is obtained as the standardized values of F(D). Drought is classified according to SPEI values: −1.0 to −1.5 is moderate drought, −1.5 to −2.0 is severe drought, and below −2.0 is extreme drought.

For the determination of a season's drought, we used the value of the SPEI of the last month of that season calculated for the duration of the season. Because more water is required during sowing and mid-season compared to harvesting, droughts only during those periods were considered. The sowing and vegetative period for wheat, for example, is 7 months; hence, a 7-month SPEI value computed at the end of April was used for analysis of wheat droughts (Table 1).

Table 1. The time spans used for standardized precipitation evapotranspiration index (SPEI) calculation for droughts estimation during different cropping seasons.

Crop	Season	Period (Months)
Wheat	October–April	7
Rice	June–September	4
Corn	April–July	5

Similarly, 4-month SPEI values estimated in September and 5-month SPEI values estimated in August were used for droughts analysis during rice and corn growing seasons, respectively as given in Table 1.

The estimated SPEI values were employed to calculate drought return periods (RPs) during cropping seasons. The following steps were used for the calculation of drought RPs.

1. The time series of droughts were generated for each cropping seasons considering SPEI < −1 as drought.
2. The years without droughts were assigned a value of zero.
3. The drought frequency analysis was conducted on the non-zero values.
4. We performed a correction using non-exceedance probability (F') to account for the number of zero values [45]:

$$F' = q + (1 - q)F$$

(3)

where F is the drought occurrence probability with consideration of non-drought years and q is the ratio of the number of non-drought years to the total study period [46].

5. The drought RP with particular severity was calculated as:

$$T(x) = \frac{1}{1 - F'(x)} \tag{4}$$

In the present study, the drought events were fitted with different distributions. The Gamma distribution was found to best fit the drought events for all seasons and was therefore selected for drought frequency analysis.

3.2. Sen's Slope Estimator

Sen's slope (Q) estimates overall slope of a time series (y) as the median of slope between all the successive sampling points [47]. Mathematically, it is represented as:

$$Q = median\left[\frac{\Delta y}{\Delta t}\right] \tag{5}$$

where Δy is the change in recorded values due to change time Δt between two successive data points in the time series.

3.3. MMK Test

Trends in climate variables are usually assessed for the quantification of the changes in climate variables. The non-parametric Mann–Kendall (MK) test [48,49], suggested by the World Meteorological Organization [50], is generally used for the evaluation of the trends in atmospheric variables. The MK test is independent of data distribution and less sensitive to missing values and is thus most commonly used for assessment of trends in different climate variables. However, the major drawback of the MK test is that it is sensitive to auto-correlated data [51,52]. Several revisions of the MK test have been completed to make it insensitive to autocorrelation [53–55]. Recent studies also revealed that long-term persistence (LTP), or a slow decay of the autocorrelation function that causes a longstanding cycle in data series, also affects the significance of the trend [56]. The LTP in climatic time series occurs due to the influence of slow climatic processes that change over time. The LTP is a part of climate; therefore, it is required to separate the natural fluctuations from an external trend to estimate the trend due to global-warming-induced climate change [1,57–61]. The recent modification of the MK test (MMK) by Hamed [62] can be used for this purpose. The MK test statistic (S) of a data series y having n data points is estimated as [48]:

$$S = \sum_{k=1}^{n-1} \cdot \sum_{i=k+1}^{n} sign(y_i - y_k) \tag{6}$$

where:

$$sign(y_i - y_k) = \begin{cases} +1 \text{ when } (y_i - y_k) > 0 \\ 0 \text{ when } (y_i - y_k) = 0 \\ -1 \text{ when } (y_i - y_k) < 0 \end{cases} \tag{7}$$

The variance of S [$Var(S)$] is estimated using Z statistics to decide trend significance:

$$Z = \begin{cases} \frac{S-1}{\sqrt{Var(S)}} \text{ when } S > 0 \\ 0 \quad \text{when } S = 0 \\ \frac{S-1}{\sqrt{Var(S)}} \text{ when } S < 0 \end{cases} \tag{8}$$

If Z is significant, the MMK test de-trends the time series [62], ranks the series (R_i), and then calculates its equivalent normal variants (Z_i) as:

$$Z_i = \phi^{-1}\left(\frac{R_i}{n+1}\right) \tag{9}$$

where ϕ^{-1} represents the inverse form of normal distribution. The Hurst coefficient (H) of the series is then derived through the maximum log likelihood function to estimate the autocorrelation function for lag l on any scale using the following equation [63]:

$$\rho_l = \frac{1}{2}\left(|l+1|^{2H} - 2|l|^{2H} + |l-1|^{2H}\right) \tag{10}$$

The significance of H is decided by the use of the first and second moments for $H = 0.5$. For significant H values, the variance of S is estimated as:

$$Var(S)^{H'} = \sum_{i<j} \cdot \sum_{k<l} \frac{2}{\pi} \sin^{-1}\left(\frac{\rho|j-i| - \rho|i-l| - \rho|j-k| + \rho|i-k|}{\sqrt{(2-2\rho|i-j|)(2-2\rho|k-l|)}}\right) \tag{11}$$

where $Var(S)^{H'}$ is the biased estimate of the variance of S, which can be removed using a correction factor (B):

$$Var(S)^{H} = Var(S)^{H'} \times B \tag{12}$$

where B is a function of H, as below:

$$B = a_0 + a_1 H + a_2 H^2 + a_3 H^3 + a_4 H^4 \tag{13}$$

where the coefficients, a_0, a_1, a_2, a_3, and a_4 are the functions of the sample size n, which can be found in Hamed [62]. The significance of the MMK test is determined using $Var(S)^{H}$ instead of $Var(S)$ in Equation (8). At a 95% confidence level, the null hypothesis of no trend is rejected if $|Z| > 1.96$. In the present study, the 95% confidence level was considered for the assessment of the significance of a trend.

4. Application Results and Analysis

4.1. Meteorological Droughts During Wheat-Growing Season

The geographical distribution of the trends in precipitation, temperature, and SPEI during sowing and vegetative (mid-season) of wheat are presented in Figure 4a–c, respectively, for recent years. The magnitude of the change estimated by the use of Sen's slope at each point is presented with a resolution of $0.5° \times 0.5°$ (original GPCC/CRU data resolution) in the maps presented in Figure 4. The significant decrease or increase in trends at the 95% confidence interval (CI) calculated by the MMK test at each of the GPCC/CRU points are shown using the symbols '+' or '−', respectively.

Figure 4a shows precipitation decreased significantly in the southwest and increased in some grids in the northeast of Afghanistan. However, increases in temperature occurred almost for the whole country, except for the southeast (Figure 4b). Increases in temperature were higher in the west—up to 0.14 °C/decade (Figure 4c). The counts of the grid points where precipitation, temperature, and SPEI changed significantly considering a 50-year sliding window and a 10-year interval during 1901–2010 for the wheat cropping season are presented in Figure 5a–c, respectively. SPEI was found to be sensitive to rainfall and temperature. Decreases in SPEI were observed at up to 45 grid points (16% area) during the period temperatures were increasing at about 40 grid points (14% area) and when rainfall was decreasing at 70 grid points (25% area) until 1931. The estimated RPs of wheat drought at the different grids of Afghanistan are depicted using box plots in Figure 6a–c for three different drought severities. A longer box size or a longer whisker in the plot is an indication of a wide range of return periods of droughts. Shorter boxes or shorter whiskers indicate the proximity of the RPs of droughts to the median value. The RPs of moderate, severe, and extreme wheat droughts ranges between 5.7 and 6.8,

12.2 and 16.0, and 20.0 and 38.0 years, respectively. The median RPs were found within 6.2–6.4 for moderate, 12.7–13 for severe, and 24–25 for extreme drought classes.

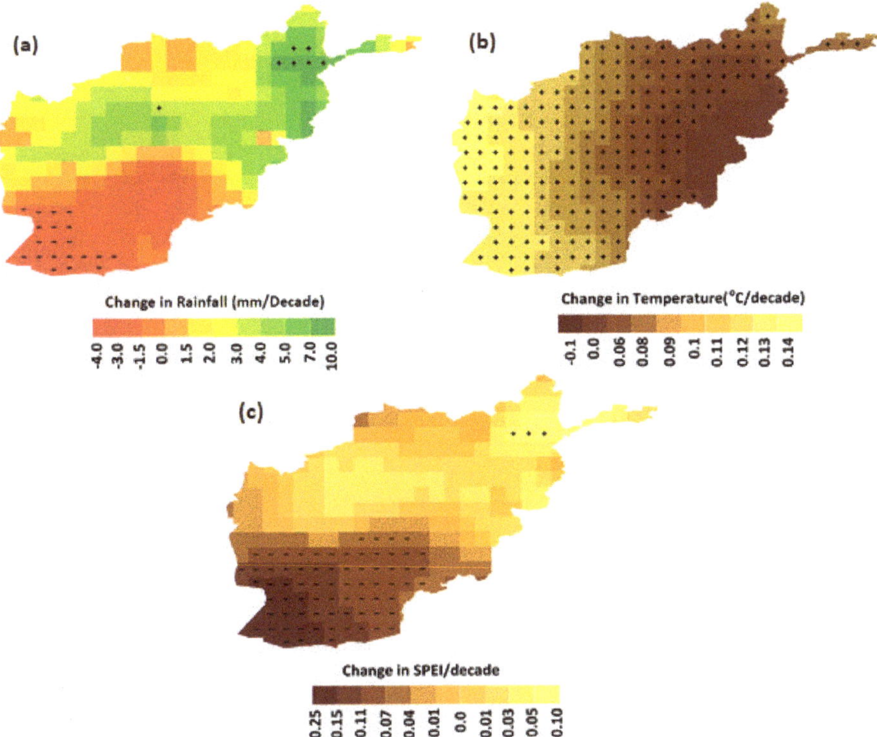

Figure 4. Trends in (**a**) precipitation, (**b**) mean temperature, and (**c**) SPEI during the wheat-growing period in Afghanistan during 1901–2010.

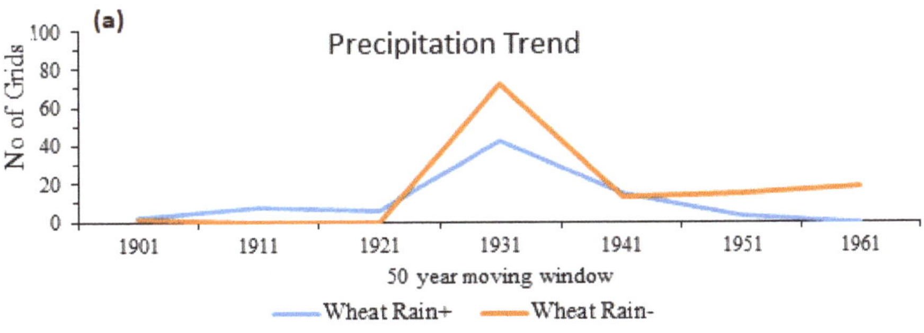

Figure 5. *Cont.*

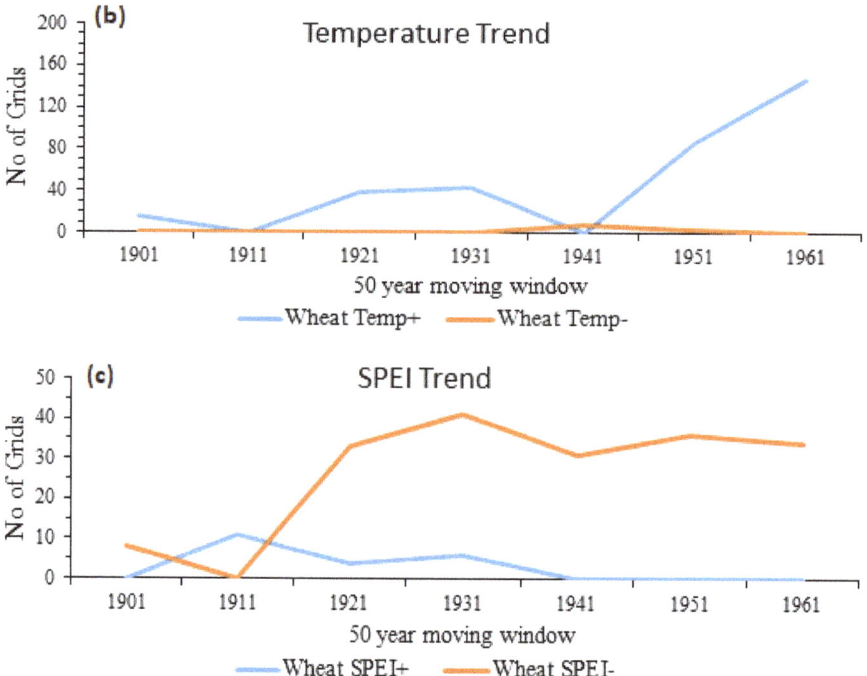

Figure 5. The number of grid points at which significant changes in (**a**) precipitation, (**b**) mean temperature, and (**c**) SPEI during wheat-growing season for different 50-year periods for 1901–2010.

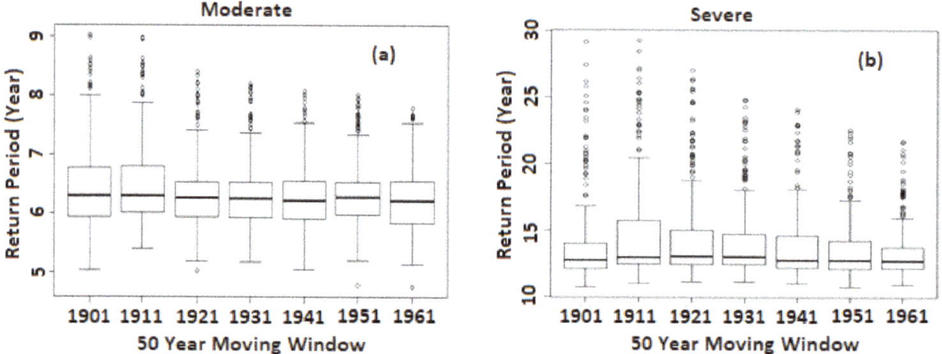

Figure 6. *Cont.*

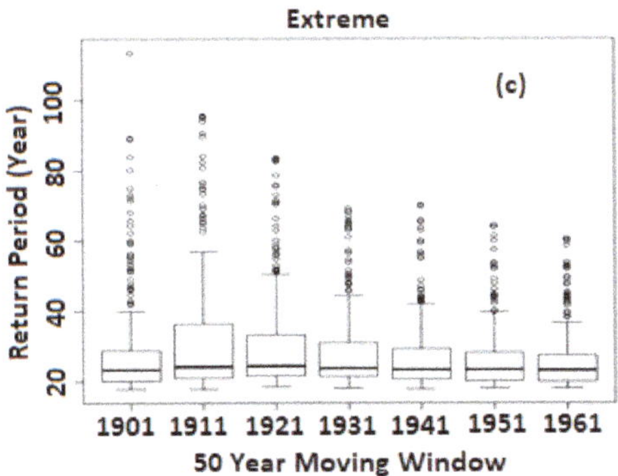

Figure 6. Box-plots showing the return periods of (**a**) moderate, (**b**) severe, and (**c**) extreme meteorological droughts during wheat-growing season observed for different 50-year windows over 1901–2010.

4.2. Meteorological Droughts in Rice-Growing Season

The geographical distributions of the changes in precipitation, temperature, and SPEI during the rice growing season are presented in Figure 7a–c respectively. Figure 7a shows a significant decrease in precipitation in the southwest and northwest and a significant increase at some grid points in the northeast and southeast. Temperature increased significantly at almost all grid points at the northwest and southwest of the country (Figure 7b). Significant decreases in temperature were observed in the southeast of the country, which caused decreases in the SPEI in the region (Figure 7c).

The grid point numbers with significant rainfall, temperature, and SPEI changes for a 50-year sliding window and a 10-year interval during 1901–2010 for the rice cropping season are outlined in Figure 8a–c, respectively. After an increase in the grid point numbers where rainfall was decreasing until 1921, there was a continuous decrease in the grid point numbers where it was decreasing after 1921. The count of grid points where precipitation was increasing sharply increased after mid-century. The drought return periods for rice are presented in Figure 9a–c for the three different drought severities. The moderate, severe, and extreme drought RPs for this crop were found to be 5.9–7.1, 12.5–15.2, and 20.0–37.0 years, respectively. The median RP values for different categories of droughts are found within 6.2–6.5 for moderate, 13.0–13.5 for severe, and 22.0–23.5 for extreme drought classes.

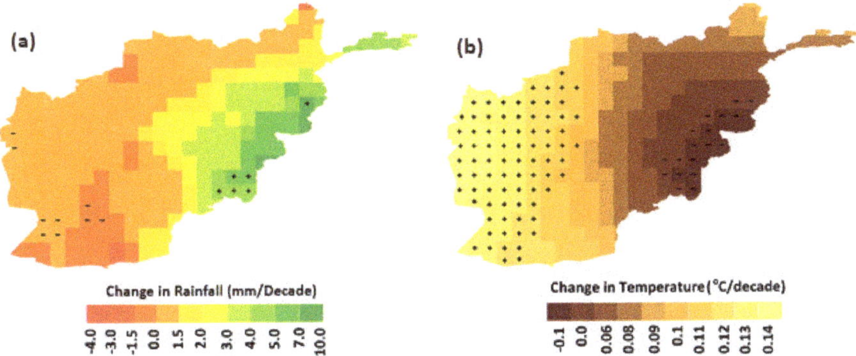

Figure 7. *Cont.*

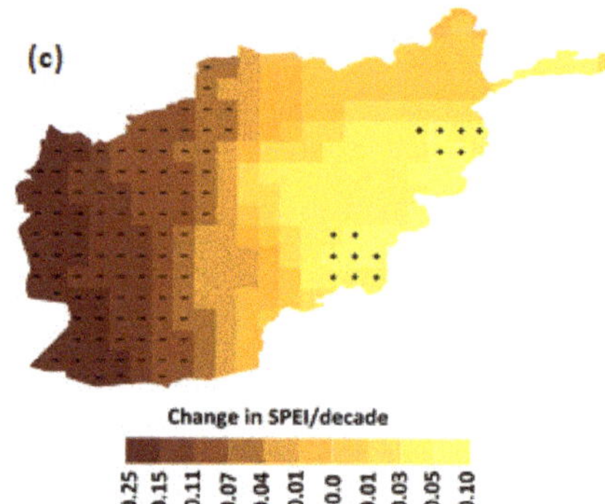

Figure 7. Trends in (**a**) precipitation, (**b**) mean temperature, and (**c**) SPEI during rice-growing period in Afghanistan during 1901–2010.

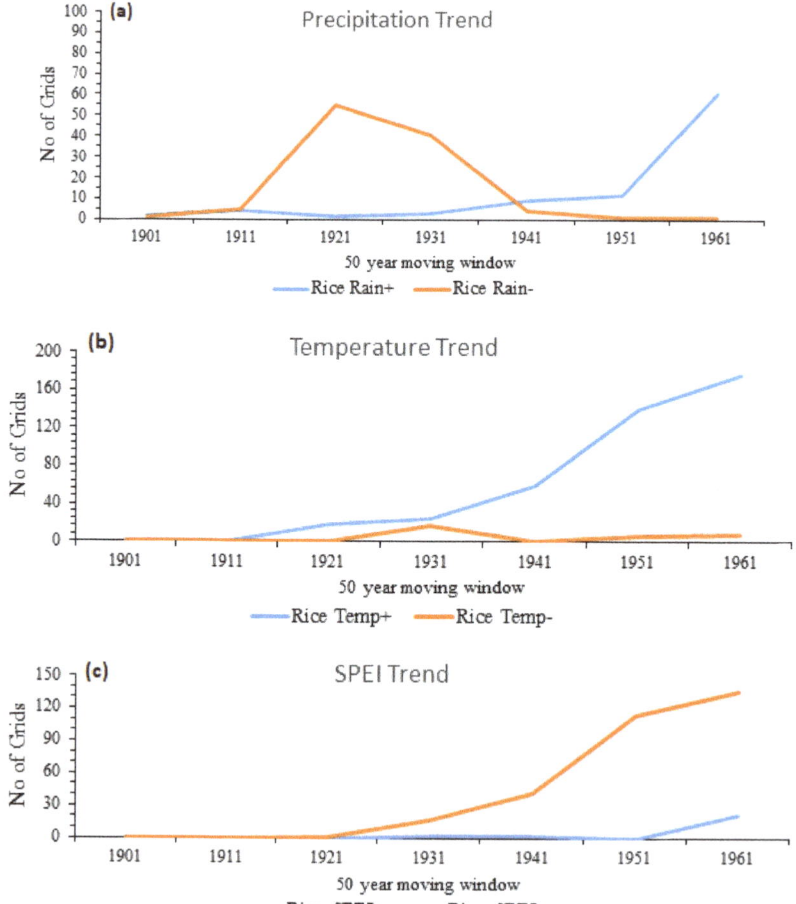

Figure 8. The number of grid points at which significant changes in (**a**) precipitation, (**b**) mean temperature, and (**c**) SPEI during rice-growing season for different 50-year windows during 1901–2010.

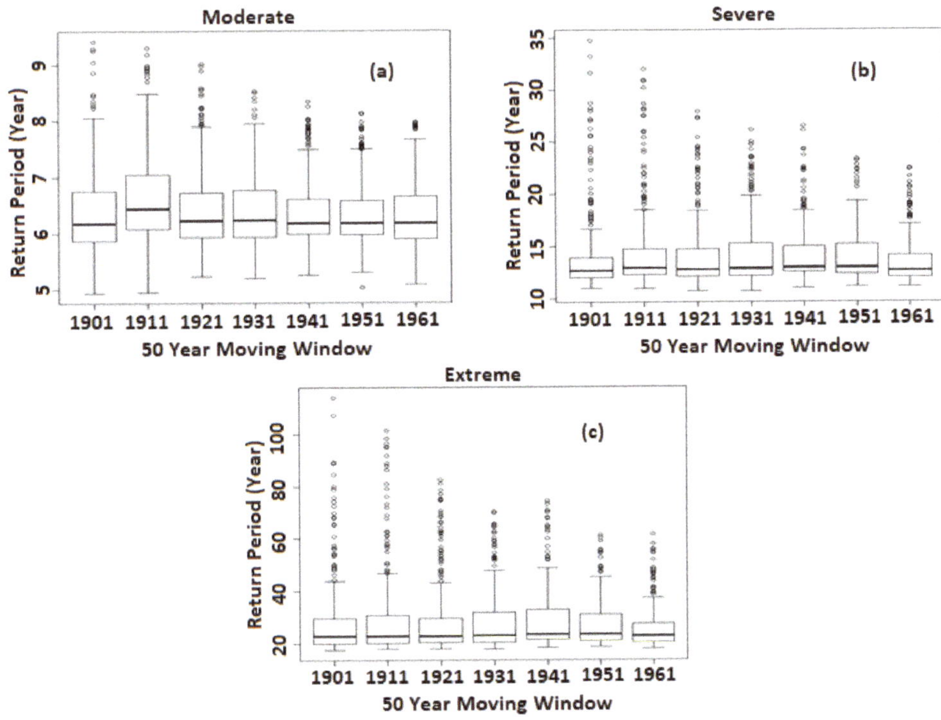

Figure 9. Box-plots showing the return period of (**a**) moderate, (**b**) severe, and (**c**) extreme droughts during rice-growing season for different 50-year windows during 1901–2010.

4.3. Meteorological Droughts During Corn-Growing Season

The geographical distributions of the changes in precipitation, temperature, and SPEI during the corn-growing season are presented in Figure 10a–c, respectively. Figure 10a shows rainfall decreased significantly at a few locations in the southwest and northwest, while significant increases occurred in the northeast and southeast of the country. Temperature increased significantly at almost all grid points in the northwest and southwest of the country (Figure 10b). Significant decreases in temperature were observed in the southeast and part of the northeast of the country. Significant increases in temperature in the northwest and southwest of the country resulted in a decrease in the SPEI in these areas (Figure 10c). The number of grids that showed significant rainfall, temperature, and SPEI changes for a 50-year sliding window and a 10-year interval during 1901–2010 for the corn cropping season are presented in Figure 11a–c, respectively. The figures show an increase in the number of grids where rainfall is increasing significantly after mid-century. The number of grids where rainfall decreased significantly also gradually increased from 1941. Drought return periods for corn are presented in Figure 12 for the three drought severities. The moderate, severe, and extreme drought return periods for this crop were 5.8–6.6, 12.0–14.0, and 19.0–27.0 years, respectively. The RP median values for the different categories of droughts were 6.2–6.3 for moderate, 12.5–13.0 for severe, and 21.0–22.0 for extreme droughts.

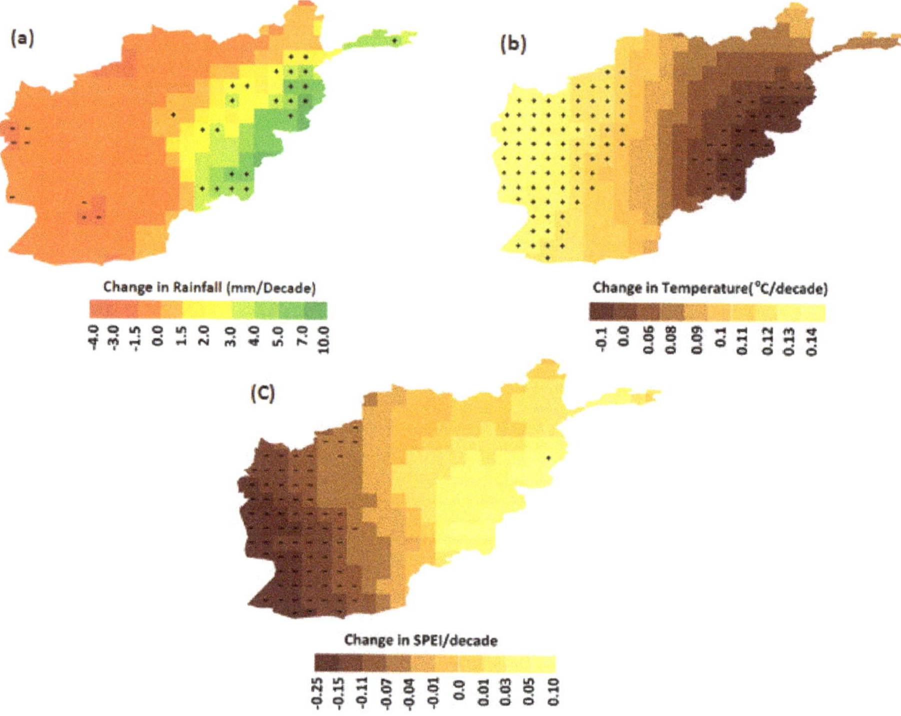

Figure 10. Geographical distributions of the changes in (**a**) precipitation, (**b**) mean temperature, and (**c**) SPEI during corn-growing season in Afghanistan during 1901–2010.

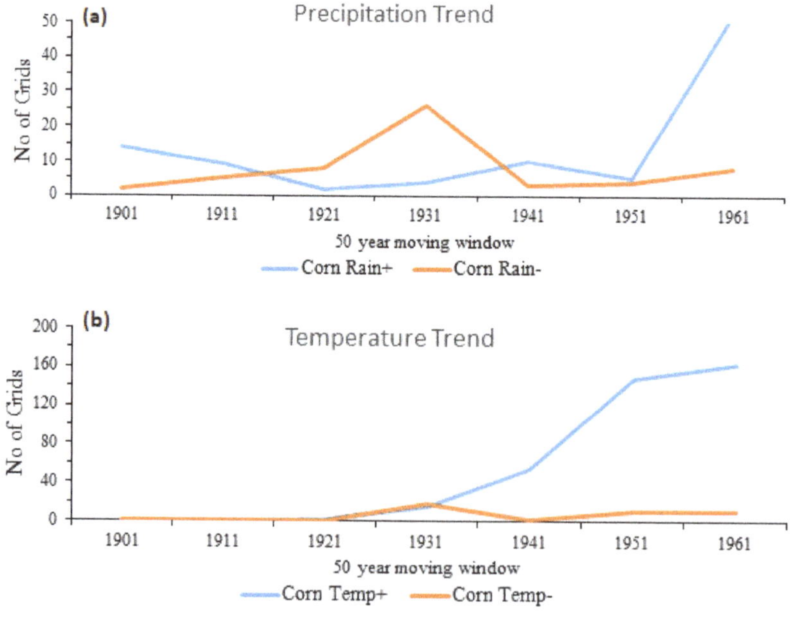

Figure 11. *Cont.*

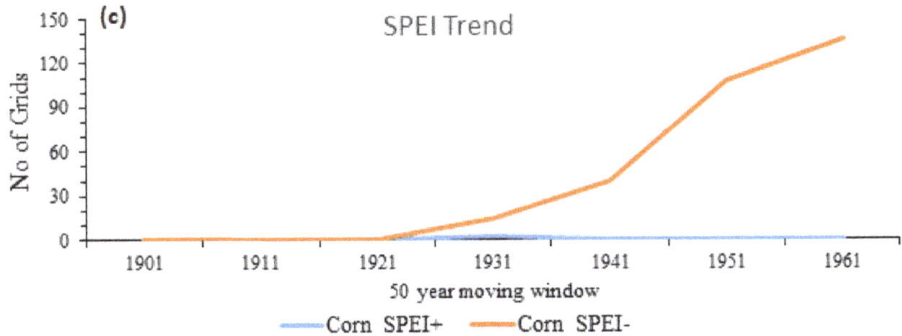

Figure 11. The number of grid points at which significant changes in (**a**) precipitation, (**b**) mean temperature, and (**c**) SPEI during corn-growing season for different 50-year windows over 1901–2010.

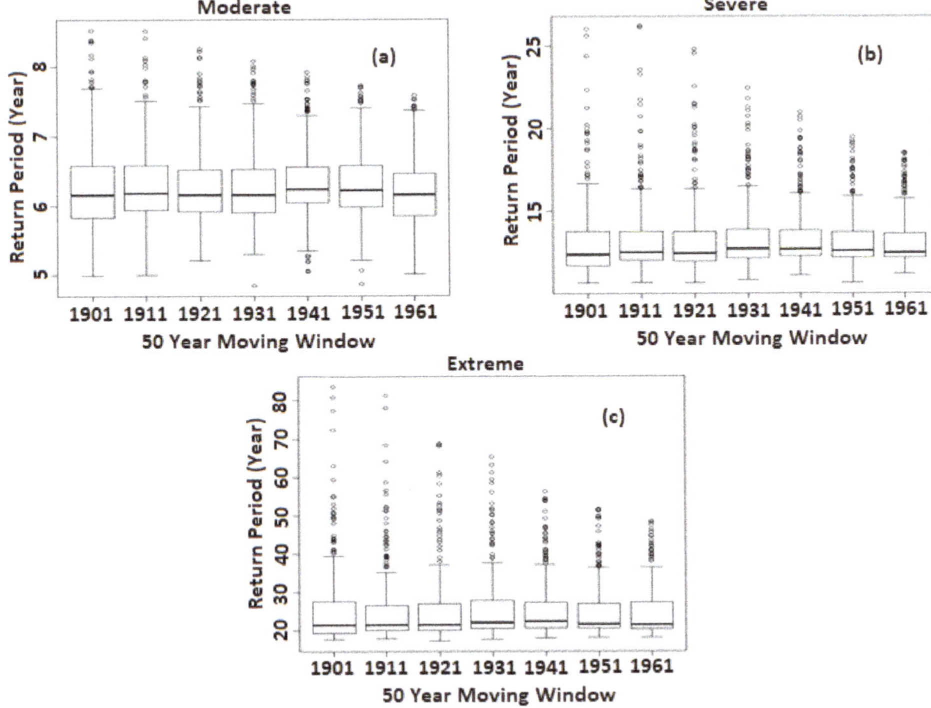

Figure 12. The box-plots showing the return periods of (**a**) moderate, (**b**) severe, and (**c**) extreme corn droughts estimated for different 50-year windows over the period 1901–2010.

5. Discussion

The changing drought characteristics during wheat-, rice-, and corn-growing periods at 281 grid points in Afghanistan during 1901–2010 using gauge-based climatic data were assessed in this study. The fitting of the SPEI values with the best PDF was used in the estimation of drought RP. The secular climate and drought index changes were assessed using the MMK test. For understanding the influences of the climatic variables on droughts, drought temporal variations and their relationships with temperature and precipitation changes were evaluated using a 50-year sliding window with 10-year intervals.

The geographical assessment of the spatial patterns of droughts for the different growing seasons showed that decreases in rainfall and increases in temperature during the wheat-growing season resulted in a decrease in the SPEI in the southwest, whereas an increase in rainfall and relatively lower rate of increase in temperature caused SPEI to rise at a few locations in the northeast. This indicates that the combined effect of rainfall and temperature played an important role in decreasing the SPEI in the southwest of Afghanistan during wheat-growing season. For rice-growing season, increases in temperature in the northwest and southwest of the country caused a decrease in the SPEI in the regions. This shows that temperature played a significant role in decreasing the SPEI in the area during rice-cropping season. In the northeast and the southeast where few grid points showed increases in precipitation and decreases in temperature, the SPEI was observed to decrease during the rice cropping season. In the corn-growing season, the significant increase in temperature in the northwest and southwest of the country resulted in a decrease in the SPEI in these areas. This is an indication that temperature was crucial to decreasing the SPEI in these areas during the corn cropping season. In the northeast and southeast parts where precipitation increased at some grids, there were no significant changes in SPEI, except for a significant increase at one grid point.

The count of grid points where rainfall, temperature, and SPEI changed for the different crops showed that during the wheat-growing season, increases were not observed in the SPEI for the study period except at the beginning of the century. A sharp increase from 1941 was observed for temperature. In recent years (1961–2010), precipitation was found to be gradually increasing with the increase in temperature in many parts of Afghanistan. This has resulted in a decreasing trend in the SPEI. Rice-growing season showed that temperature continuously increased at about 170 grid points (60% of the total grid points) until the end of the century. Continuous increases in temperature resulted in continuous decreases in the SPEI at many grid points (150 grids, about 53%) during the rice cropping season. During corn-growing season, temperature increased continuously from 1921 until the end of the study period at about 170 grid points (60% of the total study area), while there were no observable decreases in temperature at any grid point during most of the study period. The SPEI decreased continuously at 140 grid points (50% of the study area) due to continuous increases in temperature.

The return periods during different crop growing seasons showed that the box plots generally contracted over time for all categories of wheat drought. This was observed both for the median and the whiskers of the box plots. Though fluctuations were observed in the RPs during the first 50 years of the century during the rice-growing season, the return periods generally decreased for all drought classes, especially for the severe and the extreme droughts after that. Similar to rice, the corn-growing season showed some fluctuations in the drought RPs, but a generally decreasing trend was observed in the RPs of all classes of drought. The return periods for all the crops suggested that droughts occurred more frequently in areas where they were not frequent before.

Although the impacts of climate change on droughts have been widely assessed in many parts of the globe, including in Afghanistan and surrounding regions, no assessments of droughts have been conducted during the cropping seasons. In Pakistan, Ahmed et al. [26] assessed the impacts of climate variability and change on seasonal drought characteristics during two cropping seasons: Rabi and Kharif. The study showed increased drought severity during both cropping seasons, mostly in the arid and semi-arid regions of the country. Increasing droughts were mostly influenced by increasing temperature, whereas decreases in droughts were influenced by rainfall. Ta et al. [27] used SPEI, the Mann–Kendal trend test, and empirical orthogonal function analysis, and showed that drought frequency was over 42.87% in the entire Central Asia region, with Uzbekistan and Turkmenistan as the driest countries in the region. The results corroborate the findings from this study of high dryness over Afghanistan, like the neighboring countries in Central Asia.

6. Conclusions

In this study, we assessed the changes in meteorological droughts during crop growing seasons in Afghanistan. A comparison of the changing characteristics of droughts for different crop growing seasons suggested that the southern parts of the country have been strongly affected by climatic changes during wheat cropping season. The northwest and the southwest of the country were more affected by droughts due to alternations in rainfall and temperature. Afghanistan, like many other semi-arid and arid regions, is prone to the impacts of a variable climate as water is a scarce resource in such regions due to the amounts of rainfall received annually and excessive evaporation due to high temperatures. Sustainable agricultural practices are crucial for Afghanistan, where many of the populace are engaged in farming as a source of livelihood. Understanding the influences of the modification of key climate variables, like precipitation and temperature, on the changing characteristics of drought is crucial to planning reliable adaptation and mitigation drought measures under a changing climate. This is especially important during crop growing seasons because droughts can be devastating when they coincide with the crop growing season. To this effect, we revealed the changing characteristics of droughts during the wheat, rice, and corn cropping seasons under a changing climate. It is anticipated that the methods applied here can be used elsewhere and be applied in the development of adaptation and mitigation measures under a changing climate.

Author Contributions: Conceptualization, S.S.; Data curation, I.Q.; Formal analysis, I.Q.; Investigation, I.Q.; Methodology, I.Q.; Project administration, Z.M.Y.; Software, K.A.; Supervision, N.A.-A., Z.M.Y., S.S. and X.W.; Visualization, A.S. and Z.M.Y.; Writing—original draft, M.S.S., K.A. and Z.M.Y.; Writing—review & editing, A.S. and Z.M.Y.

Funding: This research received no external funding.

Conflicts of Interest: The authors declare no conflict of interest.

References

1. Salman, S.A.; Shahid, S.; Ismail, T.; Chung, E.-S.; Al-Abadi, A.M. Long-term trends in daily temperature extremes in Iraq. *Atmos. Res.* **2017**. [CrossRef]
2. Khan, N.; Shahid, S.; Juneng, L.; Ahmed, K.; Ismail, T.; Nawaz, N. Prediction of heat waves in Pakistan using quantile regression forests. *Atmos. Res.* **2019**. [CrossRef]
3. Ahmed, K.; Shahid, S.; Sachindra, D.A.; Nawaz, N.; Chung, E.S. Fidelity assessment of general circulation model simulated precipitation and temperature over Pakistan using a feature selection method. *J. Hydrol.* **2019**, *573*, 281–298. [CrossRef]
4. Mishra, A.K.; Singh, V.P. Drought modeling—A review. *J. Hydrol.* **2011**, *403*, 157–175. [CrossRef]
5. Ahmed, K.; Shahid, S.; Harun, S.b.; Wang, X.-j. Characterization of seasonal droughts in Balochistan Province, Pakistan. *Stoch. Environ. Res. Risk Assess.* **2016**. [CrossRef]
6. Shiru, M.S.; Shahid, S.; Chung, E.S.; Alias, N. Changing characteristics of meteorological droughts in Nigeria during 1901–2010. *Atmos. Res.* **2019**. [CrossRef]
7. Ahmed, K.; Shahid, S.; Ismail, T.; Nawaz, N.; Wang, X.J. Absolute homogeneity assessment of precipitation time series in an arid region of Pakistan. *Atmosfera* **2018**. [CrossRef]
8. Farooq, M.; Wahid, A.; Kobayashi, N.; Fujita, D.; Basra, S.M.A. Plant drought stress: Effects, mechanisms and management. In *Sustainable Agriculture*; Springer: Dordrecht, The Netherlands, 2009; ISBN 9789048126651.
9. Samarah, N.H. Effects of drought stress on growth and yield of barley. *Agron. Sustain. Dev.* **2005**. [CrossRef]
10. Lake, P.S. Ecological effects of perturbation by drought in flowing waters. *Freshw. Biol.* **2003**. [CrossRef]
11. Trenberth, K.E.; Dai, A.; Van Der Schrier, G.; Jones, P.D.; Barichivich, J.; Briffa, K.R.; Sheffield, J. Global warming and changes in drought. *Nat. Clim. Chang.* **2014**, *4*, 17–22. [CrossRef]
12. European Commission; European Commission—High Level Group on Science Education; European Commission—Science, Economy and Society. *Science Education Now: A Renewed Pedagogy for the Future of Europe*; Office for Official Publications of the European Communities: Luxembourg, 2007; ISBN 927905659X.
13. Wilhite, D.A.; Svoboda, M.D.; Hayes, M.J. Understanding the complex impacts of drought: A key to enhancing drought mitigation and preparedness. *Water Resour. Manag.* **2007**. [CrossRef]
14. Wilhite, D.A.; Sivakumar, M.V.K.; Pulwarty, R. Managing drought risk in a changing climate: The role of national drought policy. *Weather Clim. Extrem.* **2014**, *3*, 4–13. [CrossRef]
15. Thurman, J.N. Oklahoma in grip of new Dust Bowl. *Christ. Sci. Monit.* **1998**. Available online: https://www.csmonitor.com/1998/0824/082498.us.us.3.html (accessed on 18 April 2019).

16. Kalaugher, E.; Beukes, P.; Bornman, J.F.; Clark, A.; Campbell, D.I. Modelling farm-level adaptation of temperate, pasture-based dairy farms to climate change. *Agric. Syst.* **2017**, *153*, 53–68. [CrossRef]

17. Dale, V.H.; Joyce, L.A.; Mcnulty, S.; Neilson, R.P.; Ayres, M.P.; Flannigan, M.D.; Hanson, P.J.; Irland, L.C.; Lugo, A.E.; Peterson, C.J.; et al. Climate change and forest disturbances. *BioScience* **2001**. [CrossRef]

18. Miyan, M.A. Droughts in asian least developed countries: Vulnerability and sustainability. *Weather Clim. Extrem.* **2015**. [CrossRef]

19. Rezaeianzadeh, M.; Stein, A.; Cox, J.P. Drought forecasting using Markov Chain model and artificial neural networks. *Water Resour. Manag.* **2016**, *30*, 2245–2259. [CrossRef]

20. Reddy, C.S.; Saranya, K.R.L. Earth observation data for assessment of nationwide land cover and long-term deforestation in Afghanistan. *Glob. Planet. Chang.* **2017**, *155*, 155–164. [CrossRef]

21. Atef, S.S.; Sadeqinazhad, F.; Farjaad, F.; Amatya, D.M. Water conflict management and cooperation between Afghanistan and Pakistan. *J. Hydrol.* **2019**, *570*, 875–892. [CrossRef]

22. Muhammad, A.; Kumar Jha, S.; Rasmussen, P.F. Drought characterization for a snow-dominated region of Afghanistan. *J. Hydrol. Eng.* **2017**. [CrossRef]

23. Iqbal, M.W.; Donjadee, S.; Kwanyuen, B.; Liu, S.-y. Farmers' perceptions of and adaptations to drought in Herat Province, Afghanistan. *J. Mt. Sci.* **2018**. [CrossRef]

24. Williams, D.B. Finding Water in the Heart of Darkness: Afghanistan's Ongoing Water Challenges, 2009. Available online: https://www.earthmagazine.org/article/finding-water-heart-darkness-afghanistans-onoing-water-challenges (accessed on 24 March 2019).

25. Sorg, A.; Huss, M.; Rohrer, M.; Stoffel, M. The days of plenty might soon be over in glacierized Central Asian catchments. *Environ. Res. Lett.* **2014**, *9*, 104018. [CrossRef]

26. Ahmed, K.; Shahid, S.; Nawaz, N. Impacts of climate variability and change on seasonal drought characteristics of Pakistan. *Atmos. Res.* **2018**, *214*, 364–374. [CrossRef]

27. Ta, Z. Analysis of the spatio-temporal patterns of dry and wet conditions in Central Asia. *Atmopshere* **2018**, *1*, 7. [CrossRef]

28. Li, Z.; Chen, Y.N.; Li, W.H.; Deng, H.J.; Fang, G.H. Potential impacts of climate change on vegetation dynamics in Central Asia. *J. Geophys. Res.* **2015**, *120*, 12345–12356. [CrossRef]

29. Zoljoodi, M.; Didevarasl, A. Evaluation of spatial-temporal variability of drought events in Iran using palmer drought severity index and its principal factors (through 1951–2005). *Atmos. Clim. Sci.* **2013**. [CrossRef]

30. Shroder, J.F. *Natural Resources in Afghanistan: Geographic and Geologic Perspectives on Centuries of Conflict*; Elsevier: San Diego, CA, USA, 2014; ISBN 0128005459.

31. Thomas, V.; Ramzi, A.M. SRI contributions to rice production dealing with water management constraints in northeastern Afghanistan. *Paddy Water Environ.* **2011**, *9*, 101–109. [CrossRef]

32. Vicente-Serrano, S.M.; Beguería, S.; López-Moreno, J.I. A multiscalar drought index sensitive to global warming: The standardized precipitation evapotranspiration index. *J. Clim.* **2010**, *23*, 1696–1718. [CrossRef]

33. Becker, A.; Finger, P.; Meyer-Christoffer, A.; Rudolf, B.; Schamm, K.; Schneider, U.; Ziese, M. A description of the global land-surface precipitation data products of the global precipitation climatology centre with sample applications including centennial (trend) analysis from 1901–present. *Earth Syst. Sci. Data* **2013**, *5*, 71–99. [CrossRef]

34. Harris, I.; Jones, P.D.; Osborn, T.J.; Lister, D.H. Updated high-resolution grids of monthly climatic observations—The CRU TS3.10 Dataset. *Int. J. Clim.* **2014**, *34*, 623–642. [CrossRef]

35. Haritashya, U.K.; Bishop, M.P.; Shroder, J.F.; Bush, A.B.G.; Bulley, H.N.N. Space-based assessment of glacier fluctuations in the Wakhan Pamir, Afghanistan. *Clim. Chang.* **2009**, *94*, 5–18. [CrossRef]

36. Palka, E.J. *Afghanistan: A Regional Geography*; United States Military Academy, West Point Department of Geography & Environmental Engineering: West Point, NY, USA, 2001.

37. Aich, V.; Akhundzadah, N.; Knuerr, A.; Khoshbeen, A.; Hattermann, F.; Paeth, H.; Scanlon, A.; Paton, E. Climate change in Afghanistan deduced from reanalysis and coordinated regional climate downscaling experiment (CORDEX)—South Asia simulations. *Climate* **2017**, *5*, 38. [CrossRef]

38. Shiru, M.S.; Shahid, S.; Chung, E.-S.; Alias, N.; Scherer, L. A MCDM-based framework for selection of general circulation models and projection of spatio-temporal rainfall changes: A case study of Nigeria. *Atmos. Res.* **2019**. [CrossRef]

39. Kishore, P.; Jyothi, S.; Basha, G.; Rao, S.V.B.; Rajeevan, M.; Velicogna, I.; Sutterley, T.C. Precipitation climatology over India: Validation with observations and reanalysis datasets and spatial trends. *Clim. Dyn.* **2016**, *46*, 541–556. [CrossRef]

40. Schneider, U.; Becker, A.; Finger, P.; Meyer-Christoffer, A.; Ziese, M.; Rudolf, B. GPCC's new land surface precipitation climatology based on quality-controlled in situ data and its role in quantifying the global water cycle. *Theor. Appl. Clim.* **2014**, *115*, 15–40. [CrossRef]

41. Spinoni, J.; Naumann, G.; Carrao, H.; Barbosa, P.; Vogt, J. World drought frequency, duration, and severity for 1951–2010. *Int. J. Clim.* **2014**, *34*, 2792–2804. [CrossRef]

42. Tuklimat, N.N.A.; Harun, S.; Shahid, S. Comparison of different methods in estimating potential évapotranspiration at Muda Irrigation Scheme of Malaysia. *J. Agric. Rural Dev. Trop. Subtrop.* **2012**, *113*, 77–85.

43. Stagge, J.H.; Tallaksen, L.M.; Xu, C.-Y.; Van Lanen, H.A.J. Standardized precipitation-evapotranspiration index (SPEI): Sensitivity to potential evapotranspiration model and parameters. In *Hydrology in a Changing World: Environmental and Human Dimensions, Proceedings of the Flow Regimes from International Experimental and Network Data (FRIEND)-Water 2014, Montpellier, France, 7–10 October 2014*; International Association of Hydrological Sciences (IAHS) Publisher: Oxfordshire, UK, 2014.

44. Beguería, S.; Vicente-Serrano, S.M.; Reig, F.; Latorre, B. Standardized precipitation evapotranspiration index (SPEI) revisited: Parameter fitting, evapotranspiration models, tools, datasets and drought monitoring. *Int. J. Climatol.* **2014**, *34*, 3001–3023. [CrossRef]

45. Condon, L.E.; Gangopadhyay, S.; Pruitt, T. Climate change and non-stationary flood risk for the Upper Truckee River Basin. *Hydrol. Earth Syst. Sci. Discuss.* **2014**, *11*, 5077–5114. [CrossRef]

46. Mohsenipour, M.; Shahid, S.; Chung, E.-s.; Wang, X.-j. Changing pattern of droughts during cropping seasons of Bangladesh. *Water Resour. Manag.* **2018**. [CrossRef]

47. Sen, P.K. Estimates of the regression coefficient based on Kendall's tau. *J. Am. Stat. Assoc.* **1968**, *63*, 1379–1389. [CrossRef]

48. Mann, H.B. Nonparametric tests against trend. *Econometrica* **1945**. [CrossRef]

49. Kendall, M.G. *Rank Correlation Methods*; Griffin: Oxford, UK, 1948.

50. World Meteorological Organization (WMO). *Analyzing Long Time Series of Hydrological Data with Respect to Climate Variability*; WCAP-3, WMO/TD-No: 224; World Meteorological Organization: Geneva, Switzerland, 1988; pp. 1–12.

51. Ahmed, K.; Shahid, S.; Othman, R.; Harun, S.B.; Wang, X.-j. Evaluation of the performance of gridded precipitation products over Balochistan Province, Pakistan. *Desalin. Water Treat.* **2017**. [CrossRef]

52. Khan, N.; Shahid, S.; Ismail, T.; Wang, X.-J. Spatial distribution of unidirectional trends in temperature and temperature extremes in Pakistan. *Theor. Appl. Climatol.* **2018**. [CrossRef]

53. Hamed, K.H.; Rao, A.R. A modified Mann-Kendall trend test for autocorrelated data. *J. Hydrol.* **1998**, *204*, 182–196. [CrossRef]

54. Yue, S.; Pilon, P.; Cavadias, G. Power of the Mann–Kendall and Spearman's rho tests for detecting monotonic trends in hydrological series. *J. Hydrol.* **2002**, *259*, 254–271. [CrossRef]

55. Yue, S.; Wang, C. The Mann-Kendall test modified by effective sample size to detect trend in serially correlated hydrological series. *Water Resour. Manag.* **2004**, *18*, 201–218. [CrossRef]

56. Koutsoyiannis, D.; Montanari, A. Statistical analysis of hydroclimatic time series: Uncertainty and insights. *Water Resour. Res.* **2007**, *43*, W05429. [CrossRef]

57. Nashwan, M.S.; Shahid, S.; Abd-Rahim, N. Unidirectional trends in annual and seasonal climate and extremes in Egypt. *Theor. Appl. Climatol.* **2018**. [CrossRef]

58. Salman, S.A.; Shahid, S.; Ismail, T.; Ahmed, K.; Wang, X.-J. Selection of climate models for projection of spatiotemporal changes in temperature of Iraq with uncertainties. *Atmos. Res.* **2018**, *213*, 509–522. [CrossRef]

59. Nashwan, M.S.; Shahid, S.; Wang, X.-J. Uncertainty in estimated trends using gridded rainfall data: A case study of Bangladesh. *Water* **2019**, *11*, 349. [CrossRef]

60. Khan, N.; Pour, S.H.; Shahid, S.; Ismail, T.; Ahmed, K.; Chung, E.-S.; Nawaz, N.; Wang, X. Spatial distribution of secular trends in rainfall indices of peninsular Malaysia in the presence of long-term persistence. *Meteorol. Appl.* **2019**. [CrossRef]

61. Pour, H.S.; Abd-Wahab, A.K.; Shahid, S.; Wang, X. Spatial pattern of the unidirectional trends in thermal bioclimatic indicators in Iran. *Sustainability* **2019**, *11*, 2287. [CrossRef]

62. Hamed, K.H. Trend detection in hydrologic data: The Mann-Kendall trend test under the scaling hypothesis. *J. Hydrol.* **2008**, *349*, 350–363. [CrossRef]
63. Koutsoyiannis, D.; Koutsoyiannis, D. Climate change, the Hurst phenomenon, and hydrological statistics. *Hydrol. Sci. J.* **2003**, *48*. [CrossRef]

Article

Hydrologic Risk Assessment of Future Extreme Drought in South Korea Using Bivariate Frequency Analysis

Ji Eun Kim [1], Jiyoung Yoo [2], Gun Hui Chung [3] and Tae-Woong Kim [2,*]

[1] Department of Civil and Environmental System Engineering, Hanyang University, Seoul 04763, Korea; helloje2@hanyang.ac.kr

[2] Department of Civil and Environmental Engineering, Hanyang University, Ansan 15588, Korea; jyyoo84@gmail.com

[3] Department of Civil Engineering, Hoseo University, Asan 31499, Korea; gunhui@hoseo.edu

* Correspondence: twkim72@hanyang.ac.kr; Tel.: +82-31-400-5184

Received: 28 August 2019; Accepted: 29 September 2019; Published: 30 September 2019

Abstract: Recently, climate change has increased the frequency of extreme weather events. In South Korea, extreme droughts are frequent and cause serious damage. To identify the risk of extreme drought, we need to calculate the hydrologic risk using probabilistic analysis methods. In particular, future hydrologic risk of extreme drought should be compared to that of the control period. Therefore, this study quantitatively assessed the future hydrologic risk of extreme drought in South Korea according to climate change scenarios based on the representative concentration pathway (RCP) 8.5. A threshold level method was applied to observation-based rainfall data and climate change scenario-based future rainfall data to identify drought events and extract drought characteristics. A bivariate frequency analysis was then performed to estimate the return period considering both duration and severity. The estimated return periods were used to calculate and compare hydrologic risks between the control period and the future. Results indicate that the average duration of drought events for the future was similar with that for the control period, however, the average severity increased in most future scenarios. In addition, there was decreased risk of maximum drought events in the Yeongsan River basin in the future, while there was increased risk in the Nakdong River basin. The median of risk of extreme drought in the future was calculated to be larger than that of the maximum drought in the control period.

Keywords: bivariate frequency analysis; climate change; drought; hydrologic risk

1. Introduction

Climate change induces changes of rainfall pattern as the consequence of global warming. Contrary to climate variability, climate change is characterized by a slow-varying oscillation in the climate system and can modify the possibility of extreme events by changing hydrological cycles [1]. According to the precipitation outlook based on climate change scenarios [2–4], climate change is likely to produce very different patterns of flood and drought around the world [1]. Compared with other natural disasters, the spatial extent of drought is usually much larger and the duration is greater [5]. In addition, the frequency of extreme droughts will increase and the damage caused by extreme droughts will be severe. Therefore, drought should be quantitatively analyzed to mitigate damage. Due to the multi-sided characteristics and the non-structural impacts of drought, it is difficult to quantify drought damage [6].

For quantitative drought analysis, drought should be defined and characterized. Various drought indices related to agricultural, soil moisture and streamflow data are used to quantify the climate dryness anomalies. For example, Standardized Precipitation Index (SPI) and Palmer Hydrological Drought Index (PHSI) have been used in practice to quantitatively analyze the characteristics of meteorological and hydrological droughts [7,8]. A threshold level method is another frequently applied tool for defining a drought event and has practical advantages—(i) no a-priori knowledge of probability distributions is required and (ii) it simply and directly produces drought characteristics such as duration and severity [9,10]. Mirakbari et al. [11] and Yu et al. [12] defined drought events using a certain threshold level of hydro-meteorological variables and analyzed the probabilistic return period through hydrological drought frequency analysis.

To estimate the risk of extreme drought, the drought frequency analysis is usually performed first. In this study, the drought frequency analysis was conducted using drought characteristics calculated by the threshold level method. Because droughts are characterized by highly correlated random variables, such as duration, intensity and peak, their joint and conditional distributions, as well as their marginal distributions, need to be analyzed to understand multivariate behaviors. Using joint and conditional distributions, the limitations of the univariate frequency analysis could be overcome and the multivariate behavior of droughts could be understood more thoroughly [13]. Both drought duration and severity could be determined easily and simultaneously through bivariate frequency analysis. Since drought duration and severity may be represented by different marginal distributions, copulas can be used to construct the joint distribution function of drought characteristics [14]. For example, Yoo et al. [15] generated synthetic monthly rainfall using the copula-GARCH rainfall generation model, after investigating performance of various copula functions. Janga and Ganguli [5] presented a copula-based multivariate probabilistic approach to analyze the severity–duration–frequency (S-D-F) relationship of drought events in western Rajasthan, India.

Since the hydrologic risk of drought is defined as the probability that a given level of dryness is exceeded in any year, results of bivariate frequency analysis can be used to analyze the risk for certain drought events. Ganguli and Janga [16] presented a risk assessment of hydrologic extreme droughts using copula-based distribution functions in the Saurashtra and Kutch regions of Gujarat, India. Recently, Yoo et al. [17] compared and analyzed drought risks based on the SPI and the Standardized Precipitation Evapotranspiration Index (SPEI) using future climate change scenario data. Kim et al. [18] performed temporal and spatial analyses on extreme drought events on the Korean peninsula to estimate hydrologic risk and evaluate future droughts using future climate change scenario data. Yu et al. [19] conducted a bivariate frequency analysis using copula functions to determine critical severity according to the non-exceedance probability and calculated the hydrologic risk for extreme drought events.

Although numerous studies have focused on drought risk analysis using observed data, more studies about hydrologic risk are needed to quantitatively analyze future extreme droughts. It also needs to compare and analyze the risk of extreme drought for future periods in South Korea under various climate change scenarios. Therefore, this study estimates the hydrologic risk for extreme drought in South Korea and investigates the changes of the drought risk in future.

2. Study area and Data

In this study, drought characteristics were analyzed for 109 medium-sized watersheds in South Korea, as shown in Figure 1. The drought characteristics of medium-sized watersheds directly contribute to regional drought management according to the governmental natural disaster response system.

Observed daily precipitation data (January 1983–December 2014) were obtained from the Korea Meteorological Administration (KMA) and climate change scenario-based daily precipitation data (January 2015–December 2099) were obtained from the Asia-Pacific Economic Cooperation Climate Center (APCC). The observed data were analyzed for a control period and the climate change scenario-based data were analyzed for the future period based on the Representative Concentration Pathway (RCP) 8.5 scenario. Although there are four RCP scenarios such as 2.6, 4.5, 6.0 and 8.5 according to greenhouse gas emissions, the RCP 8.5 represents increases in emissions based on the current trend, without a greenhouse gas reduction policy [20]. The RCP 8.5 scenario is more appropriate for extreme drought analysis than other RCP scenarios, because it realizes the extreme circumstances. 19 Global Climate Models (GCMs) were selected to provide future precipitation data, as shown in in Table 1.

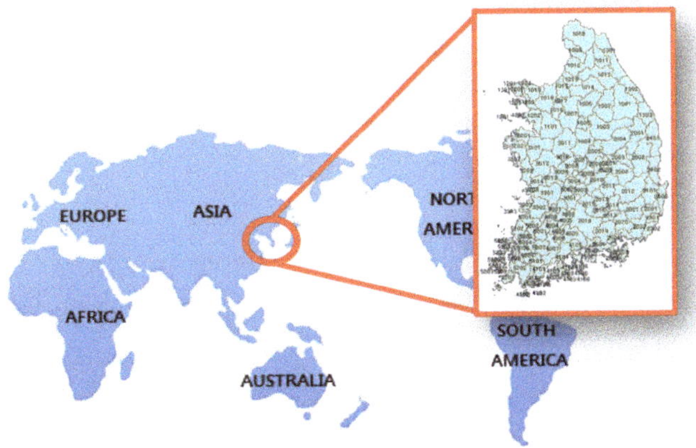

Figure 1. Map of the study area.

Table 1. Dataset of global climate models used in this study.

Model	Institution
CanESM2	Canadian Centre for Climate Modelling and Analysis
CCSM4	National Center for Atmospheric Research
CESM1-BGC	National Center for Atmospheric Research
CESM1-CAM5	National Center for Atmospheric Research
CMCC-CM	Centro Euro-Mediterraneo per I Cambiamenti Climatici
CMCC-CMS	Centro Euro-Mediterraneo per I Cambiamenti Climatici
CNRM-CM5	Centre National de Recherches Meteorologiques
GFDL-ESM2G	Geophysical Fluid Dynamics Laboratory
GFDL-ESM2M	Geophysical Fluid Dynamics Laboratory
HadGEM2-AO	Met Office Hadley Centre
HadGEM2-ES	Met Office Hadley Centre
INM-CM4	Institute for Numerical Mathematics
IPSL-CM5A-LR	Institut Pierre-Simon Laplace
IPSL-CM5A-MR	Institut Pierre-Simon Laplace
MIROC5	Atmosphere and Ocean Research Institute
MPI-ESM-LR	Max Planck Institute for Meteorology
MPI-ESM-MR	Max Planck Institute for Meteorology
MRI-CGCM3	Meteorological Research Institute
NorESM1-M	Norwegian Climate Centre

Direct use of GCMs in regional-scale hydrologic analysis is constrained, because the atmospheric dynamics at scales greater than 2 degrees of longitude and latitude yields unrealistic hydrologic outcomes [21]. Dynamic and statistical methods are used to downscale the output from GCMs to regional-scale. The dynamic method is a physical process to enhance the spatial resolution of GCMs climate output through regional climate models (RCMs) [22]; however, it results in biases in precipitation simulations from RCMs due to limited understanding of regional climate processes [1]. The climate change scenario-based data used in this study were provided by the APCC after correcting the bias of GCMs using statistical relationships between observed historical climate data and GCM-based climate variables [22]. In addition, the KMA certifies and releases climate change scenario-based data after validating and examining their soundness and applicability.

3. Methodology

3.1. Threshold Level Method

The most widely used method to define drought is the threshold level method (TLM). The TLM is based on run theory and has two main advantages—(i) no a priori knowledge of probability distributions is required and (ii) it directly produces drought characteristics if the threshold is set according to drought-impacted sectors [9]. Because the TLM is convenient for identifying drought events, it is efficient for applying hydro-meteorological time series compared to other drought identification methods using a drought index like SPI and PDSI [23]. Wada et al. [24] applied the TLM to quantify the impact of human water consumption on intensity and frequency of hydrological drought worldwide over the period 1960–2010. Yoo et al. [25] employed the 3-month and 12-month moving averages as the threshold level and analyzed drought characteristics. Due to the simplicity and efficiency for defining drought, the TLM was used in this study to analyze the drought risk effectively. According to the TLM, the period during which rainfall is less than the predefined threshold is regarded as the drought duration and the accumulated sum of rainfall deficit volume during the duration is defined as the drought severity [26,27]. The period from the beginning of any drought event to the beginning of the next drought event is called interarrival time. Figure 2 illustrates the details of defining drought characteristics by applying the monthly threshold level.

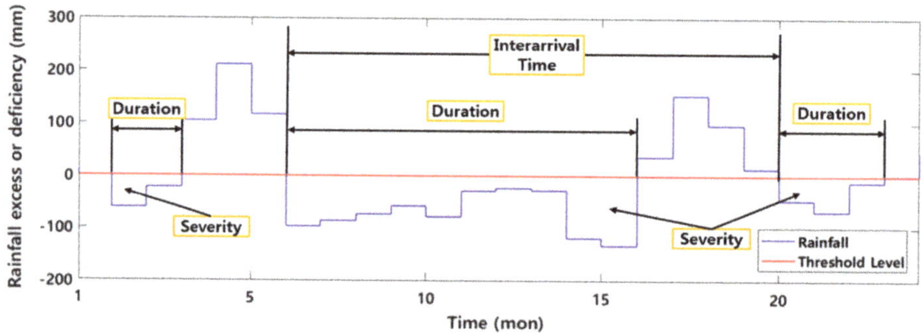

Figure 2. Definition of drought characteristics using a threshold level.

There are many ways to determine the threshold level. Because the characteristics of drought vary according to threshold level, an appropriate threshold level should be chosen [12]. In this study, the threshold level was determined as percentiles of precipitation based on the empirical cumulative distribution function (ECDF). The threshold level commonly ranges between the 70th and 95th percentiles depending on regional characteristics when applying the TLM to precipitation [28,29]. The threshold may be constant or a variable over a year. If the threshold is derived from the ECDF, it implies that the whole precipitation record (or a predefined period) is used in the derivation of the threshold [30]. This leads to increased accuracy of the rainfall deficit investigation during months with low and high precipitation [31].

3.2. Bivariate Frequency Analysis

To quantify the hydrologic risk of drought, it is necessary to estimate the return period through drought frequency analysis. Drought frequency analysis is divided into univariate and bivariate frequency analysis. Since drought is usually represented by its duration and severity, univariate frequency analysis using a single drought property may not reveal significant relationships among the drought properties [10,32]. Therefore, both duration and severity should be considered together in bivariate frequency analysis to reveal the multivariate characteristics of drought [32]. There are several kinds of bivariate drought frequency analysis—(i) parametric methods, (ii) nonparametric methods and (iii) copula function methods. Parametric methods fit an assumed theoretical probability distribution for drought properties, whereas nonparametric methods do not require any assumptions about probability distribution [26]. Lately, copula function methods have been widely applied in hydrology and water resources engineering [33–40] because they have an advantage of reflecting the dependency structure between two variables and combining their probability distributions [41–44]. The copula function for calculating the joint probability distribution of drought duration (D) and severity (S) is shown in Equation (1).

$$F_{D,S}(d,s) = C(F_D(d), F_S(s)) \tag{1}$$

where $F_{D,S}(d,s)$ is the bivariate distribution function of two correlated random variables D and S. $F_D(d)$ and $F_S(s)$ are marginal CDFs of D and S, respectively. C represents a copula function.

Copula functions are classified into four categories—Archimedean, elliptical, extreme value and Plackett. The Archimedean copula can be constructed easily, includes a large variety of copula functions and can be applied whether the correlations amongst hydrologic variables are positive or negative [19,45,46]. Therefore, the Archimedean copula functions are more applicable to drought frequency analysis compared to the others. Table 2 shows the Archimedean copula functions.

Table 2. Equations of Archimedean copula functions.

Name	Equation $C(u,v)$	Note
Clayton	$\left(u^{-\theta} + v^{-\theta} - 1\right)^{-1/\theta}$	
Frank	$\frac{1}{\theta} \ln\left[1 + \frac{[e^{-\theta u}-1][e^{-\theta v}-1]}{e^{-\theta}-1}\right]$	u and v denote random variates θ is a parameter.
Gumbel	$\exp\left[-\left\{\{-\ln u\}^{\theta} + \{-\ln v\}^{\theta}\right\}^{1/\theta}\right]$	

The return period of an event can be calculated as the inverse of the exceedance probability of the event. The univariate frequency analysis needs only marginal probability of interest variable [47], while bivariate frequency analysis considers two distinctive joint probabilities—(i) AND probability that two variables are exceed specific values, (ii) OR probability that one of two variables is exceeded specific values. In design of a water resources system, it is appropriate to consider the AND probability for a safe design [48]. The joint return period of drought is available for the AND probability, as given by Equation (2).

$$T_{DS} = \frac{E(L)}{P(D \geq d \text{ and } S \geq s)} = \frac{E(L)}{1 - F_D(d) - F_S(s) + C(F_D(d), F_S(s))} \tag{2}$$

where E(L) is the expected drought interarrival time.

There are many ways to assess the risk of drought—it is commonly divided into drought risk in the field of natural disasters and drought risk in the fields of hydrology and water resources engineering. Drought risk in the natural disaster field is defined as the scale of potential damage caused by a disaster for a specific period in a particular region and is calculated as the product of hazard and vulnerability. Drought risk in the fields of hydrology and water resources engineering is represented by the hydrologic risk (R) of failure which is the probability of occurrence when the magnitude of the design return period (T_{DS}) is exceeded within the expected life (n) of the structure [49], as given by Equation (3).

$$R = 1 - \left(1 - \frac{1}{T_{DS}}\right)^n \tag{3}$$

The hydrologic risk for each drought event can be estimated through the bivariate return period calculated by Equation (2).

4. Results

4.1. Definition of a Drought Event

In this study, to extract drought characteristics from precipitation time series, the TLM was applied. The threshold level was set based on the percentile of the empirical cumulative distribution function obtained by ordering precipitation data. The threshold level was determined as the closest percentile, within the 70–95th percentiles, which generated as many drought events as using the SPI. After applying the determined threshold to the observed and climate change scenario-based precipitation, drought events were extracted and the duration and severity of drought events were calculated. For the efficiency of analysis, the characteristics of drought events in 109 watersheds are summarized in Table 3, in which average duration and severity are shown for five large river basins.

Table 3. Characteristics of drought events.

Characteristics	Dataset		Han River	Nakdong River	Geum River	Seomjin River	Yeongsan River
Average Duration (mon)	Observed data		3.42	3.56	3.41	3.47	3.43
	GCMs	Lowest	3.18 (HadGEM2-AO)	2.97 (HadGEM2-AO)	3.19 (HadGEM2-AO)	3.23 (IPSL-CM5A-MR)	3.21 (HadGEM2-AO)
		Highest	3.73 (CFDL-ESM2M)	3.83 (CMCC-CM)	3.72 (CMCC-CM)	3.73 (IPSL-CM5A-LR)	3.87 (CFDL-ESM2M)
		Average	3.44	3.51	3.50	3.51	3.47
Average Severity (mm)	Observed data		660.8	600.3	601.8	711.2	630.9
	GCMs	Lowest	599.9 (CESM1-BGC)	503 (CESM1-BGC)	558.5 (CESM1-BGC)	654.7 (CESM1-BGC)	579.2 (CESM1-BGC)
		Highest	804.1 (INM-CM4)	725.4 (HadGEM2-AO)	771.8 (HadGEM2-AO)	832.7 (CNRM-CM5)	742.3 (CFDL-ESM2M)
		Average	685.73	6225.73	661.41	760.20	669.51

Table 3 demonstrates that HadGEM2-ES produced the lowest average duration and the differences between average durations from the observed and the future scenario were small. CESM1-BGC produced the lowest average severity and the averages of future severity were larger than those of control period for all large basins. It was expected that there will be many droughts in the future with similar duration but greater severity than in the control period. These findings suggest that extreme droughts occur in the future, consistent with results of previous studies.

4.2. Drought Risk Analysis Using Bivariate Frequency Analysis

Various distributions such as exponential, normal, gamma, lognormal, Poisson, Weibull, generalized extreme value and generalized pareto were employed as possible distributions for duration and severity. After estimating distribution parameters using the maximum likelihood method, the optimal probabilistic distributions were individually determined for 109 medium-sized watershed based on χ^2 goodness-of-fit test. Among the Archimedean copula functions such as Clayton, Frank and Gumbel, a suitable copula function was selected for observed data in individual watershed and applied to the future climate change scenario data to perform the bivariate frequency analysis. Consequently, the largest value of the bivariate return periods was chosen as the maximum drought event for a watershed and dataset. The characteristics of maximum drought events are shown in Figures 3 and 4. Table 4 summaries the characteristics of maximum drought event.

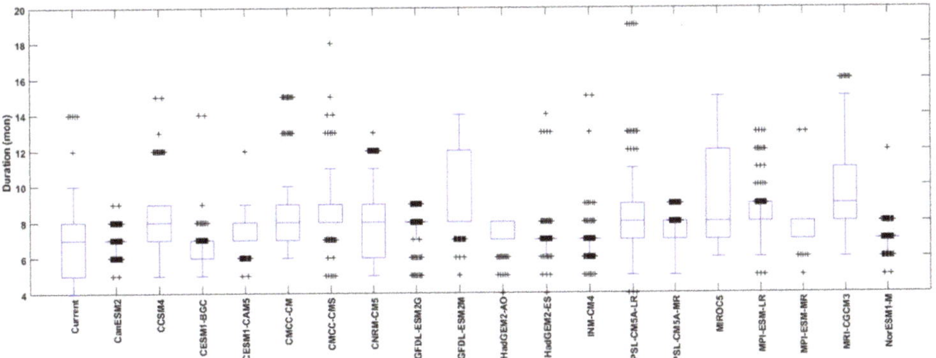

Figure 3. Duration of maximum droughts for each dataset.

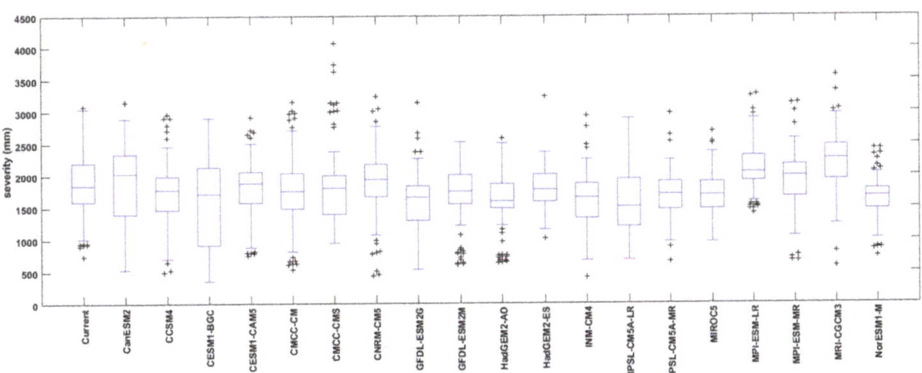

Figure 4. Severity of maximum droughts for each dataset.

Table 4. Characteristics of maximum drought events.

Characteristics	Dataset		Han River	Nakdong River	Geum River	Sumjin River	Yeongsan River
	Observed data		12	14	14	14	14
Average Duration (mon)	GCMs	Lowest	8 (CESM1-BGC)	7 (HadGEM2-AO)	9 (GFDL-ESM2M)	7 (HadGEM2-AO)	10 (CESM1-BGC)
		Highest	25 (MPI-ESM-MR)	25 (MPI-ESM-MR)	25 (MPI-ESM-MR)	19 (CESM1-CAM5)	29 (MPI-ESM-MR)
		Average	13.2	11.9	14.1	12.0	13.8
	Observed data		2673.7	2127.6	2745.7	2238.8	2307.5
Average Severity (mm)	GCMs	Lowest	2169.9 (CESM1-BGC)	1474.1 (CESM1-BGC)	2526.3 (CMCC-CM)	2020.7 (HadGEM2-AO)	2073.1 (NorESM1-M)
		Highest	5081.9 (MPI-ESM-MR)	3964.9 (MPI-ESM-MR)	4793.6 (CanESM2)	3823.5 (IPSL-CM5A-LR)	4920.2 (MPI-ESM-MR)
		Average	3132.62	2178.64	3278.40	2707.99	2757.99

Comparison of duration and severity of maximum drought events for each dataset indicated that the median duration of maximum drought in the control period was 7 months. The median severity of maximum drought in the control period was 1849 mm. These medians are the criteria for future risk analysis. In this study, the risk analysis was conducted in two ways using the results of bivariate frequency analysis. We calculated hydrologic risks for the maximum drought events of the control period and then estimated hydrologic risks of extreme droughts in the future period, which are the drought events that are larger than the maximum drought in the control period. The design return periods of a large dam are set to 100–200 years in South Korea. Therefore, in this study, the expected life (n in Equation (3)) of hydrologic risk was set to 100 years. The hydrologic risks of maximum drought events calculated for 109 watersheds are shown in Figure 5.

In the control period, the highest risk was in the downstream area of the Yeongsan River and the upstream area of the Nakdong River and the lowest risk was in the downstream area of the Nakdong River. However, the results of the future period indicated that the Yeongsan River basin was greatly decreased and the Nakdong River basin was rather dangerous in both upstream and downstream areas. Future drought events exceeding the maximum drought of the control period were defined as extreme droughts and were analyzed using the hydrologic risk. The results for future extreme droughts are shown in Figure 6. The red line represents the median risk for the maximum drought of the control period.

(a) Observed   (b) CanESM2   (c) CCSM4   (d) CESM1-BGC

(e) CESM1-CAM5   (f) CMCC-CM   (g) CMCC-CMS   (h) CNRM-CM5

(i) GFDL-ESM2G   (j) GFDL-ESM2M   (k) HadGEM2-AO   (l) HadGEM2-ES

Figure 5. *Cont.*

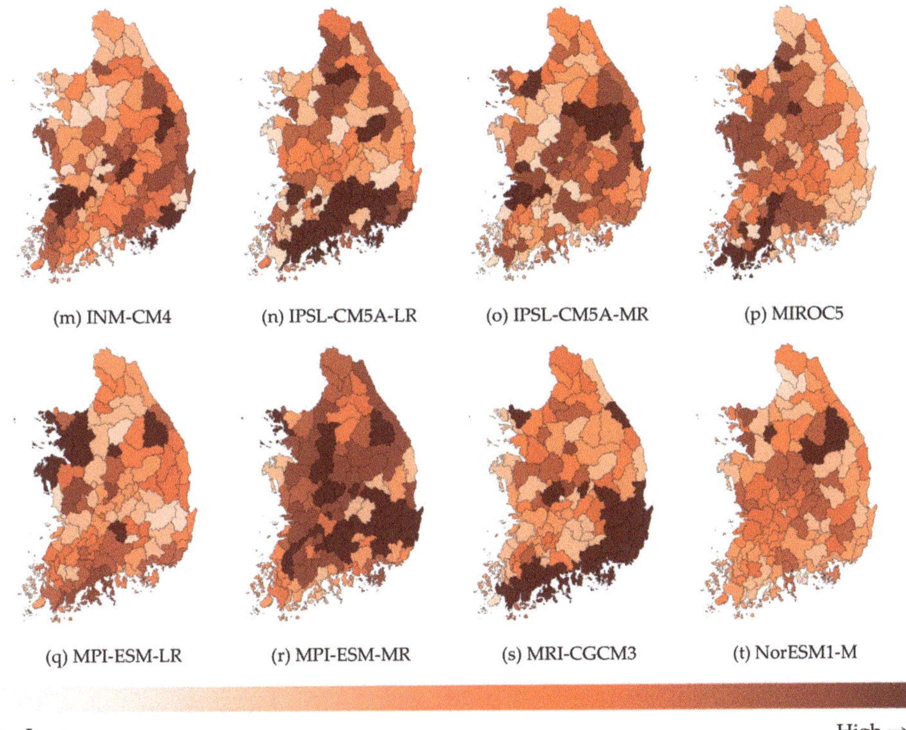

Figure 5. Hydrologic risk of maximum drought events for each dataset.

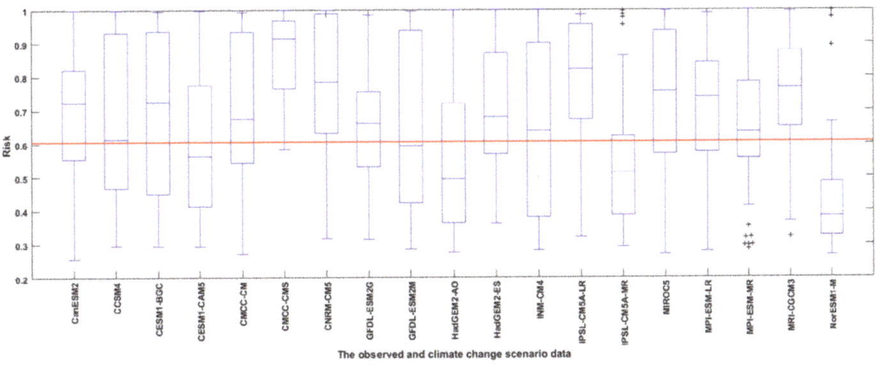

Figure 6. Hydrologic risk of extreme drought for each dataset.

The median risk of the maximum drought of the control period was 0.61, which was lower than most of the future scenario data. CMCC-CMS of the future scenario data had the highest value of 0.91 compared to the control period and NorESM1-M had the lowest value of 0.38. Table 5 demonstrates the average risk for extreme drought by watershed.

Table 5. Average hydrologic risk of extreme drought event in each basin.

Basin (River)	CanESM2	CCSM4	CESM1-BGC	CESM1-CAM5	CMCC-CM	CMCC-CMS	CNRM-CM5	GFDL-ESM2G	GFDL-ESM2M	HadGEM2-AO	HadGEM2-ES	INM-CM4	IPSL-CM5A-LR	IPSL-CM5A-MR	MIROC5	MPI-ESM-LR	MPI-ESM-MR	MRI-CGCM3	NorESM1-M
Han River	0.64	0.53	0.56	0.65	0.83	0.84	0.68	0.55	0.55	0.62	0.72	0.61	0.59	0.51	0.73	0.78	0.67	0.78	0.48
Nakdong River	0.79	0.73	0.83	0.51	0.54	0.89	0.85	0.71	0.70	0.69	0.73	0.66	0.78	0.56	0.73	0.63	0.60	0.69	0.58
Geum River	0.65	0.58	0.75	0.45	0.57	0.94	0.55	0.67	0.71	0.58	0.78	0.46	0.77	0.59	0.68	0.64	0.68	0.91	0.38
Seomjin River	0.68	0.82	0.62	0.78	0.81	0.82	0.84	0.60	0.67	0.52	0.62	0.83	0.81	0.57	0.84	0.84	0.75	0.71	-
Yeongsan River	0.26	0.54	0.66	0.63	0.65	0.80	0.80	0.52	0.68	0.41	-	0.45	0.79	0.38	0.69	0.75	0.96	0.70	0.31

These results show that CMCC-CMS had the highest risk compared to other GCMs, while NorESM1-M had the lowest risk. In addition, the risks of the Han River, Nakdong River and Seomjin River basins were expected to increase, while the risks of the Geum River and Yeongsan River basins were expected to be similar to those in the control period. Those results are similar to those of Park et al. [50] and the Nakdong, Seomjin and Han River basins are expected to experience more severe drought in the future.

5. Conclusions

In this study, we conducted a bivariate frequency analysis for 109 medium-sized watershed of South Korea using the observed data and 19 RCP 8.5-based synthetic data and then assessed the hydrologic risk of drought for control and future periods. The main conclusions are as follows; (1) The average drought duration of control period and the average drought duration of future period were similar. However, the average drought in the future was higher according to climate change. Consequently, it was expected that extreme droughts would occur in the future. (2) Considering hydrologic risk of drought, the Yeongsan River basin and the upstream area of the Nakdong River basin were dangerous in the control period. In the future, the hydrologic risk of the Yeongsan River basin decreased but that of the Nakdong River basin increased in the downstream region. (3) The hydrologic risks of extreme drought, which was more severe than the maximum drought in the control period, were up to 1.5 times higher than those of the control period. The Han River, Nakdong River and Seomjin River basins predicted high risk of severe drought.

This study demonstrated that drought would be more severe and dangerous in the future in South Korea. Current design standards do not consider the change of extreme drought risk in the future. Therefore, design standards for water resources systems should be established to cope with future extreme droughts considering the climate change. The results of this study can be useful in establishing risk-based quantitative design standards of water resource systems for drought mitigation. Consequently, the design standards of the water resource system can be changed considering the risk of drought and the government can give priority to the area with high risk when preparing the drought measures.

Author Contributions: J.E.K.: Writing-original draft; J.Y.: Investigation and validation; G.H.C.: Writing-review & editing; T.-W.K.: Conceptualization and project administration

Funding: This work was supported by Korea Environment Industry & Technology Institute (KEITI) through Advanced Water Management Research Program, funded by Korea Ministry of Environment (Grant. 83070).

Conflicts of Interest: The authors declare no conflict of interest.

References

1. Lee, J.H.; Kwon, H.H.; Jang, H.W.; Kim, T.W. Future changes in drought characteristics under extreme climate change over South Korea. *Adv. Meteorol.* **2016**, *2016*, 1–19. [CrossRef]
2. Kwon, H.-H.; Khalil, A.F.; Siegfried, T. Analysis of extreme summer rainfall using climate teleconnections and typhoon characteristics in South Korea. *J. Am. Water Resour. Assoc.* **2008**, *44*, 436–448. [CrossRef]
3. Kwon, H.-H.; Brown, C.; Lall, U. Climate informed flood frequency analysis and prediction in Montana using hierarchical Bayesian modeling. *Geophys. Res. Lett.* **2008**, *35*, 05404. [CrossRef]
4. Kwon, H.-H.; Sivakumar, B.; Moon, Y.-I.; Kim, B.-S. Assessment of change in design flood frequency under climate change using a multivariate downscaling model and a precipitation runoff model. *Stoch. Environ. Res. Risk Assess.* **2011**, *25*, 567–581. [CrossRef]
5. Janga, R.M.; Ganguli, P. Application of copulas for derivation of drought severity-duration-frequency curves. *Hydrol. Process.* **2012**, *26*, 1672–1685. [CrossRef]
6. Obasi, G.O.P. WMO's role in the international decade for natural disaster reduction. *Bull. Am. Meteorol. Soc.* **1994**, *75*, 1655–1662. [CrossRef]

7. Lee, J.H.; Park, S.Y.; Kim, J.S.; Sur, C.; Chen, J. Extreme drought hotspot analysis for adaptation to a changing climate: Assessment of applicability to the five major river basins of the Korean Peninsula. *Int. J. Climatol.* **2018**, *38*, 4025–4032. [CrossRef]

8. Waseem, M.; Ajmal, M.; Kim, T.W. Development of a new composite drought index for multivariate drought assessment. *J. Hydrol.* **2015**, *527*, 30–37. [CrossRef]

9. Van Huijgevoort, M.H.J.; Hazenberg, P.; van Lanen, H.A.J.; Uijlenhoet, R. A generic method for hydrological drought identification across different climate regions. *Hydrol. Earth Syst. Sci.* **2012**, *16*, 2437–2451. [CrossRef]

10. Shiau, J.T.; Shen, H.W. Recurrence analysis of hydrologic droughts of differing severity. *J. Water Resour. Plan. Manag.* **2001**, *127*, 30–40. [CrossRef]

11. Mirakbari, M.; Ganji, A.; Fallah, S.R. Regional bivariate frequency analysis of meteorological droughts. *J. Hydrol. Eng.* **2010**, *15*, 985–1000. [CrossRef]

12. Yu, J.S.; Shin, J.Y.; Kwon, M.S.; Kim, T.W. Bivariate drought frequency analysis to evaluate water supply capacity of multi-purpose dams. *J. Korean Soc. Civ. Eng.* **2017**, *37*, 231–238. [CrossRef]

13. Kim, T.-W.; Valdés, J.B.; Yoo, C.S. Nonparametric approach for estimating return periods of droughts in arid regions. *J. Hydrol. Eng.* **2003**, *8*, 237–246. [CrossRef]

14. Mirabbasi, R.; Fakheri-Fard, A.; Dinpashoh, Y. Bivariate drought frequency analysis using the copula method. *Theor. Appl. Climatol.* **2012**, *108*, 191–206. [CrossRef]

15. Yoo, J.; Kim, D.; Kim, H.; Kim, T.W. Application of copula functions to construct confidence intervals of bivariate drought frequency curve. *J. Hydrol. Environ. Res.* **2016**, *11*, 113–122. [CrossRef]

16. Ganguli, P.; Reddy, M.J. Risk assessment of droughts in Gujarat using bivariate copulas. *Water Resour. Manag.* **2012**, *26*, 3301–3327. [CrossRef]

17. Yoo, J.Y.; Kwon, H.H.; Lee, J.H.; Kim, T.W. Influence of evapotranspiration of future drought risk using bivariate drought frequency curves. *KSCE J. Civ. Eng.* **2016**, *20*, 2059–2069. [CrossRef]

18. Kim, N.S.; Kim, J.S.; Jang, H.W.; Lee, J.H. Hydrologic risk analysis based on extreme drought over the Korean peninsula under climate change. *J. Korea Soc. Hazard Mitig.* **2015**, *15*, 45–52. [CrossRef]

19. Yu, J.S.; Yoo, J.Y.; Lee, J.H.; Kim, T.W. Estimation of drought risk through the bivariate drought frequency analysis using copula functions. *J. Korea Water Resour. Assoc.* **2016**, *49*, 217–225. [CrossRef]

20. Park, M.W.; Lee, O.J.; Park, Y.K.; Kim, S.D. Future drought projection In Korea under AR5 RCP climate change scenarios. *J. Korea Soc. Hazard Mitig.* **2015**, *15*, 423–433. [CrossRef]

21. Wood, A.W.; Leung, L.R.; Sridhar, V.; Lettenmaier, D.P. Hydrologic implications of dynamical and statistical approaches to downscaling climate model outputs. *Clim. Chang.* **2004**, *62*, 189–216. [CrossRef]

22. Bae, D.H.; Jung, I.W.; Lettenmaier, D.P. Hydrologic uncertainties in climate change from IPCC AR4 GCM simulations of the Chungju basin, Korea. *J. Hydrol.* **2011**, *401*, 90–105. [CrossRef]

23. Kwak, J.W.; Lee, S.D.; Kim, Y.S.; Kim, J.S. Return period estimation of droughts using drought variables from standardized precipitation index. *J. Korea Water Resour. Assoc.* **2013**, *46*, 795–805. [CrossRef]

24. Wada, Y.; van Beek, L.P.H.; Wanders, N.; Bierkens, M.F.P. Human water consumption intensifies hydrological drought worldwide. *Environ. Res. Lett.* **2013**, *8*, 034036. [CrossRef]

25. Yoo, J.Y.; Kim, T.W.; Kim, S.D. Drought frequency analysis using cluster analysis and bivariate probability distribution. *J. Korea Water Resour. Assoc.* **2010**, *30*, 599–606. [CrossRef]

26. Yoo, J.Y.; Kwon, H.H.; Kim, T.W.; Ahn, J.H. Drought frequency analysis using cluster analysis and bivariate probability distribution. *J. Hydrol.* **2012**, *14*, 102–111. [CrossRef]

27. Sung, J.H.; Chung, E.S. Proposal and application of water deficit-duration-frequency curve using threshold level method. *J. Korea Water Resour. Assoc.* **2014**, *47*, 997–1005. [CrossRef]

28. Carrão, H.; Singleton, A.; Naumann, G.; Barbosa, P.; Vogt, J. An optimized system for the classification of meteorological drought intensity with applications in frequency analysis. *J. Appl. Meteor. Climatol.* **2014**, *53*, 1943–1960. [CrossRef]

29. Van Loon, A.F. Hydrological drought explained. *Wiley Interdiscip. Rev. Water* **2015**. [CrossRef]

30. Hisdal, H.; Tallaksen, T. *Drought Event Definition*; Technical Report; ARIDE Technical Report NO. 6; University of Oslo: Oslo, Norway, 2000.

31. Karimi, M.; Shahedi, K. Hydrological drought analysis of Karkheh River basin in Iran using variable threshold level method. *Curr. World Environ. J.* **2013**, *8*, 419–428. [CrossRef]

32. Kim, T.-W.; Valdés, J.B.; Yoo, C.S. Nonparametric approach for bivariate drought characterization using Palmer drought index. *J. Hydrol. Eng.* **2006**, *11*, 134–143. [CrossRef]

33. De Michele, C.; Salvadori, G. A generalized pareto intensity-duration model of storm rainfall exploiting 2-copulas. *J. Geophys. Res.* **2003**, *108*, 4067. [CrossRef]
34. De Michele, C.; Salvadori, G.; Canossi, M.; Petaccia, A.; Rosso, R. Bivariate statistical approach to check adequacy of dam spillway. *J. Hydrol. Eng.* **2005**, *10*, 50–57. [CrossRef]
35. Favre, A.C.; Adlouni, S.E.; Perreault, L.; Thiemonge, N.; Bobbe, B. Multivariate hydrological frequency using copulas. *Water Resour. Res.* **2004**, *40*, 1–12. [CrossRef]
36. Salvadori, G.; De Michele, C. Analytical calculation of storm volume statistics with pareto-like intensity-duration marginals. *Geophys. Res. Lett.* **2004**, *31*, 1–4. [CrossRef]
37. Salvadori, G.; De Michele, C. Frequency analysis via copulas: Theoretical aspects and applications to hydrological events. *Water Resour. Res.* **2004**, *40*, 1–17. [CrossRef]
38. Salvadori, G.; De Michele, C. Statistical characterization of temporal structure of storms. *Adv. Water Resour.* **2006**, *29*, 827–842. [CrossRef]
39. Salvadori, G.; De Michele, C. On the use of copulas in hydrology: Theory and practice. *J. Hydrol. Eng.* **2007**, *12*, 369–380. [CrossRef]
40. Wong, G. A comparison between the Gumbel-Hougaard and distorted Frank copulas for drought frequency analysis. *Int. J. Hydrol. Sci. Technol.* **2013**, *3*, 77–91. [CrossRef]
41. Wong, G.; Lambert, M.F.; Leonard, M.; Metcalfe, A.V. Drought analysis using trivariate copulas conditional on climate states. *J. Hydrol. Eng.* **2010**, *15*, 129–141. [CrossRef]
42. Lee, T.; Salas, J.D. Copula-based stochastic simulation of hydrological data applied to Nile River flows. *Hydrol. Res.* **2011**, *42*, 318–330. [CrossRef]
43. Yoo, J.Y.; Yu, J.S.; Kwon, H.H.; Kim, T.W. Determination of drought events considering the possibility of relieving drought and estimation of design drought severity. *J. Korea Water Resour. Assoc.* **2016**, *49*, 275–282. [CrossRef]
44. Shiau, J.-T.; Wang, H.-Y.; Tsai, C.-T. Bivariate frequency analysis of flood using copulas. *J. Am. Water Resour. Assoc.* **2006**, *42*, 1549–1564. [CrossRef]
45. Nelson, R.B. *An Introduction to Copulas*; Springer: New York, NY, USA, 1999.
46. Zhang, L.; Singh, V.P. Bivariate flood frequency analysis using the copula method. *J. Hydrol. Eng.* **2006**, *11*, 150–164. [CrossRef]
47. Chen, L.; Sinngh, V.P.; Guo, S.; Mishra, A.K.; Guo, J. Drought analysis using copulas. *J. Hydrol. Eng.* **2013**, *18*, 797–808. [CrossRef]
48. Kwon, Y.-M.; Kim, T.-W. Derived I-D-F curve in Seoul using bivariate precipitation frequency analysis. *J. Korean Soc. Civ. Eng.* **2009**, *29*, 155–162.
49. Chow, V.T.; Maidment, D.R.; Mays, L. *Applied Hydrology*; McGraw-Hill: New York, NY, USA, 1988; p. 572.
50. Park, B.S.; Lee, J.H.; Kim, C.J.; Jang, H.W. Projection of future drought of Korea based on probabilistic approach using multi-model and multi climate change scenarios. *J. Korean Soc. Civ. Eng.* **2013**, *33*, 1871–1885. [CrossRef]

Article

Changes in Future Drought with HadGEM2-AO Projections

Minsung Kwon [1] and Jang Hyun Sung [2,*]

[1]   K-water Research Institute, Water Resources Research Center, Daejeon 34045, Korea; minsung8151@gmail.com
[2]   Ministry of Environment, Han River Flood Control Office, Seoul 06501, Korea
*   Correspondence: jhsung1@korea.kr; Tel.: +82-2-590-9983

Received: 6 January 2019; Accepted: 10 February 2019; Published: 12 February 2019

Abstract: The standardized precipitation index (SPI)—a meteorological drought index—uses various reference precipitation periods. Generally, drought projections using future climate change scenarios compare reference SPIs between baseline and future climates. Here, future drought was projected based on reference precipitation under the baseline climate to quantitatively compare changes in the frequency and severity of future drought. High-resolution climate change scenarios were produced using HadGEM2-AO General Circulation Model (GCM) scenarios for Korean weather stations. Baseline and future 3-month cumulative precipitation data were fitted to gamma distribution; results showed that precipitation of future climate is more than the precipitation of the baseline climate. When future precipitation was set as that of the baseline climate instead of the future climate, results indicated that drought intensity and frequency will decrease because the non-exceedance probability for the same precipitation is larger in the baseline climate than in future climate. However, due to increases in regional precipitation variability over time, some regions with opposite trends were also identified. Therefore, it is necessary to understand baseline and future climates in a region to better design resilience strategies and mechanisms that can help cope with future drought.

Keywords: SPI; reference precipitation; reference period; climate change

1. Introduction

Drought is a natural disaster that causes widespread damage incurring severe economic, social, and environmental costs [1]. Climate change is expected to increase drought frequency and severity in the near future [2]. Studies by the Pew Research Center have shown that drought is one of the most concerning aspects of climate change [3]. Drought can be difficult to define. As such, several indexes have been used to monitor drought, and the phenomenon is often classified as either meteorological or socio-economic [4,5]. Meteorological drought typically occurs first, and is analyzed using drought indexes that calculate precipitation surpluses and deficits [6]. The Standardized Precipitation Index (SPI) suggested by McKee et al. [7] is a representative meteorological drought index used to analyze meteorological and hydrological drought; its primary advantage is its ability to consider multiple time scales [8,9]. Experts of 'The Inter-Regional Workshop on Indices and Early Warning Systems for Drought', held at the University of Nebraska-Lincoln, in the United States, agreed to use SPI as a global meteorological drought index [2]. Combining high-resolution climate change scenarios with drought indexes are a common method for producing drought projections in the context of climate change. To produce a high-resolution scenario, spatial downscaling has been performed statistically and dynamically. Statistical downscaling cannot consider non-stationarity of the climate because of its assumption that the relationship between the observations and the simulation baseline period will continue in the future. However, there is an advantage to statistical downscaling in that it can be directly compared with the baseline climate through bias corrections between observations and the model. Recently, statistical downscaling techniques have been developed to consider long-term trends of the projected data, which minimized the limitations of earlier statistical downscaling techniques [10]. Eum and Cannon [10] applied Spatial Disaggregation/Quantile Delta Mapping (SDQDM) that combined daily Bias-Correction/Spatial Disaggregation (BCSD) and Quantile Delta Mapping (QDM) to produce downscaled climate projections over South Korea. Most of the previous studies for future drought projected both drought severity and duration for each period.

SPI represents drought as a shortage relative to normal, meaning that long-term precipitation data are essential to calculate it. Furthermore, SPI is the random variable of the standard normal distribution corresponding to the non-exceedance probability of a gamma distribution. Therefore, depending on the data used to estimate the probability distribution, the non-exceedance probability of cumulative precipitation can change, which in turn alters SPI. Studies of future extreme precipitation changes have primarily used comparisons of non-exceedance probabilities and return periods based on identical precipitation normals for both the baseline and future climates [11–14]. Unlike projections of extreme precipitation changes, studies of future drought projections have been largely based on non-exceedance probabilities as opposed to a comparison of cumulative precipitation. Kim et al. [15] calculated the baseline and future Standardized Precipitation Evapotranspiration Index (SPEI) using the Representative Concentration Pathway (RCP) 8.5 scenario [16]. They projected that drought frequency will increase in the future. SPEI follows the same progression as SPI except for the input variable. SPEI uses the difference between precipitation and potential evapotranspiration ($D = P - PET$) instead of precipitation alone [17,18]. They also projected a change in the drought magnitude of SPEI -1. Lee et al. [19] evaluated future changes in the spatial distribution of drought frequency and severity using Intergovernmental Panel on Climate Change (IPCC) GCM) simulations. Kyoung et al. [20] assessed future droughts in Seoul using the Special Report on Emissions Scenarios (SRES) A2 scenario and predicted that future long-term droughts would increase in severity. Park et al. [21] projected future droughts in Korea using SRES A2 and RCP8.5 scenarios and projected that drought frequency will increase in the future. Kim et al. [22] analyzed the future drought of the Han River Basin using the RCP8.5 scenario and showed that drought frequency will increase in that location. Park et al. [23] projected the future drought in Korea using the RCP8.5 scenario and found projected increases in both drought duration and severity. In these studies, drought was compared based on each non-exceedance probability in the baseline and future climates. In other words, to compare baseline and future periods, these studies employed the same non-exceedance probabilities. The same non-exceedance probability means that a different surplus and deficit for cumulative precipitation will occur with the same frequency in baseline and future. However, the sustainability of the present drought mitigation plan for the future cannot be fully evaluated because these studies did not make a quantitative comparison.

Most studies have calculated the SPI for the future and baseline period, which it evaluated the future compared to the baseline using the future climate change scenarios as shown in references [1,3] of Figure 1. The frequency and magnitude of drought were expected to increase, because of increasing of evapotranspiration due to an increase in temperature and the decreasing number of rainy days. However, climate change scenarios generally projected an increase in precipitation in the future. We assume that the criteria classifying the average state are different from future and baseline. The definition of drought is based on the criteria (or threshold) levels. Non-parametric methods using criteria precipitation can be analyzed using the same criteria as baseline and future. The SPI known as the representative parametric methods have defined the drought as excess probability, and the baseline and the future period (that is, the sample) are different, resulting in a difference of the criteria precipitation defining the drought. Therefore, we projected future droughts based on baseline precipitation called as criteria defining drought in the baseline period (Figure 1). The excess probabilities of future precipitation were calculated from the Cumulative Distribution Functions (CDF) of the gamma in the baseline climate as shown in reference [2] of Figure 1, which the frequencies of future drought were calculated as the baseline. Section 2 introduces the methodology and data, and Section 3 compares future precipitation with baseline and future climate standards. Finally, conclusions and a discussion of results are presented in Section 4.

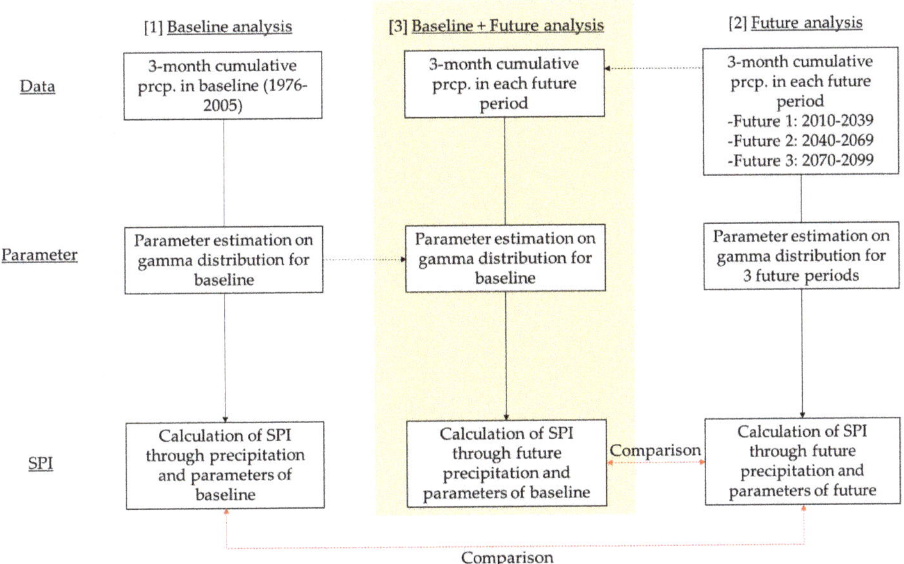

Figure 1. Concept of this Study.

2. Data and Method

2.1. Downscaling and Study Area

Evaluation periods were divided into four 30-year periods: The baseline ones (1976–2005) and three future ones (2010–2039, 2040–2069, 2070–2099). The change in Probability Density Function (PDF) of the gamma distribution was based on future climate change using baseline and future cumulative precipitation. The change in precipitation corresponding to one-month SPI (SPI−1) was projected by each of these gamma CDFs. Finally, we projected the SPI of future drought from the baseline drought criteria.

As greater than 70% of South Korea is mountainous, downscaling of climate change scenarios is essential because of large topographic impacts. Therefore, HadGEM2-AO (Hadley Centre Global Environment Model version 2 Atmosphere Ocean; [24]), a GCM based on RCP scenarios was downscaled for 57 weather stations in South Korea (Figure 2), preserving the long-term trend of the climate simulations (Figure 2). HadGEM2-AO, within the framework of CMIP5, has played an important role in assessing future climate at the national level in South Korea.

RCP 8.5 reflects the current trend of greenhouse gas (GHG) emissions, and RCP2.6 is the maximum limit at which the Earth can still have resilience. RCP4.5 and RCP6.0 are cases where a GHG reduction policy is realized to some extent. The radiative forcings of 8.5, 6.0, 4.5, and 2.6 correspond to approximately 3.6%, 2.5%, 1.9%, and 1.1% of solar radiation, respectively [25]. This study used the RCP8.5 scenario, in which the GHG emissions are relatively unchecked in the future.

Asia Pacific economic cooperation Climate Center (APCC) Integrated Modeling (AIMS) produced high-resolution data with two downscaling methods ([26]); the Simple Quantile Method (SQM) and SDQDM [10]. SDQDM preserves the long-term temporal trends and is essentially BCSD with QDM. QDM [27] (Cannon et al., 2015) is compared to synthetic data with Detrended Quantile Mapping (DQM), Burger et al. [28], which is designed to preserve any trends in the mean and with standard quantile mapping (QM). In this study, SDQDM was used as a downscaling method, and high-resolution historical simulations and future projections of daily precipitation over 30-year intervals were produced using AIMS.

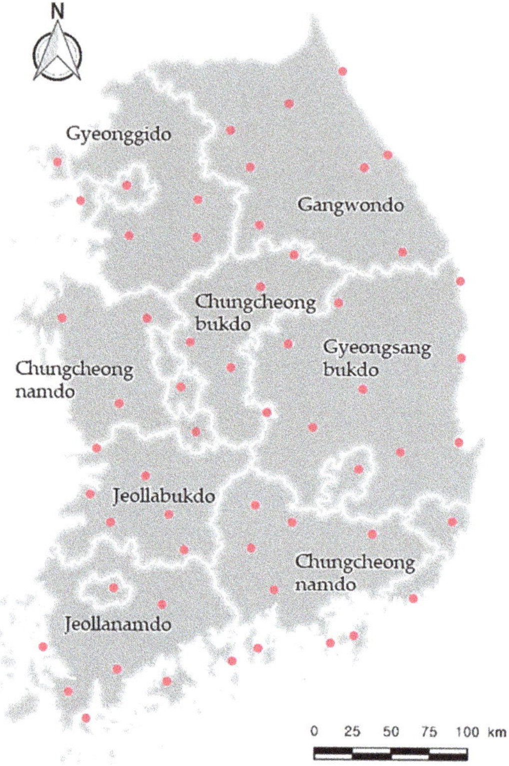

Figure 2. 57 weather stations over South Korea.

The HadGEM2-AO simulations were conducted by the National Institute of Meteorological Research/Korea Meteorological Administration (NIMR/KMA). The horizontal resolution of HadGEM2-AO was 1.875 × 1.250. HadGEM2-AO showed good performance in simulating the temperature and precipitation over Northeast Asia, particularly the Korean Peninsula, in terms of annual cycle, precipitation pattern, and timing of the rainy season ([29–31]). The HadGEM2-AO climate model became a standard scenario officially certified by the central government. Before drought projection using simulated precipitation from HadGEM2-AO, we confirmed that the simulated precipitation properly reproduced the observed precipitation to identify spatial distribution. Figure 3a shows the spatial distributions of the observed and simulated precipitation during January–March. The eastern coast and the south region had large precipitation. From April to March, the amount of precipitation in the central region was smaller than other regions in the both precipitations (Figure 3b). The simulated precipitation similar to observation in the southeast region was small during July–September (Figure 3c), but there was a difference between precipitations. From October to December, the precipitation in the eastern inland region was smaller than other regions in both precipitation (Figure 3d). We confirmed that spatial distributions and magnitude captured by the models are in good agreement with the observations.

Figure 4 shows monthly observed and simulated precipitation, which HadGEM2-AO was underestimated during the rainy season. However, we focused on employing same criteria precipitation regardless of the current and the future. Though differences were confirmed, the model projected precipitation in the current and future has been valid, therefore simulated precipitation was applied.

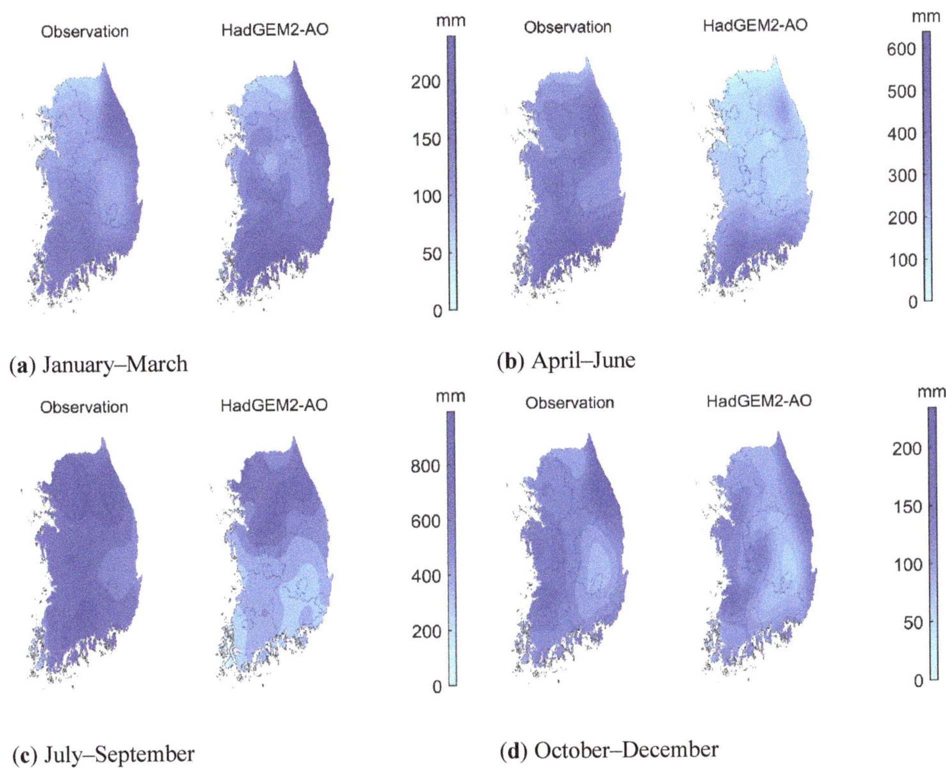

(a) January–March

(b) April–June

(c) July–September

(d) October–December

Figure 3. Spatial distribution observed and simulated 3-month cumulative precipitation.

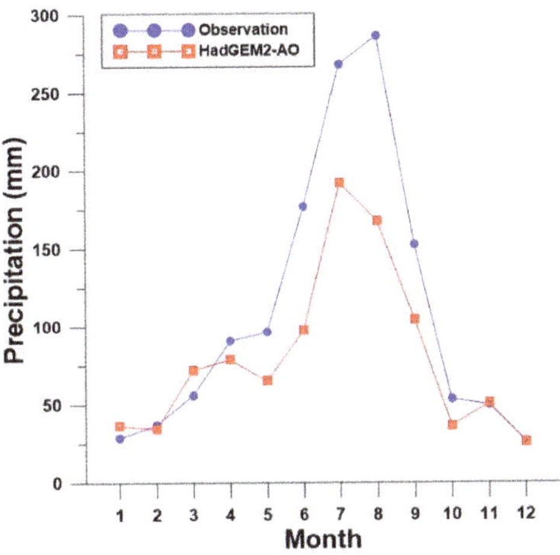

Figure 4. Monthly observed and simulated precipitation (1980–2005).

2.2. SPI

In this study, the drought was assessed using SPI and the change in precipitation corresponding to SPI of -1, which is the drought criteria in SPI that was examined. And the future drought was evaluated by applying gamma parameters of the baseline to the future precipitation. The SPI proposed by McKee et al. [7] is one of the most widely used drought indexes with the advantage of simple calculation using only precipitation. In addition, the SPI was agreed to be used by drought experts as a global meteorological drought index at the 'Inter-Regional Workshop on Indices and Early Warning Systems for Drought' held at the University of Nebraska-Lincoln in 2009 [2].

The SPI is calculated by applying a gamma distribution to the cumulative precipitation over a period, and standardizing it. The gamma distribution was not defined at zero, thus the parameters of gamma probability distribution (Equation (1)) were estimated except for the zero of the cumulative precipitation.

$$G(x) = \frac{1}{\beta^\alpha \Gamma(\alpha)} \int_0^x x^{\alpha-1} e^{-x/\beta} dx \tag{1}$$

$$H(x) = (1-q)G(x) + q \tag{2}$$

where $G(x)$ is the CDF of the gamma distribution, α and β are the shape and scale parameters, q is the probability that cumulative rainfall is zero. The SPI is calculated by substituting the probability and applying the probability (Equation (2)) to Equation (3). In Equation (3), F^{-1} is the inverse of the standard normal distribution $(F(x) = \frac{1}{\sqrt{2\pi}} \int_{-\infty}^x e^{-x^2/2})$. That is, as shown in Figure 5, SPI is the random variable (②) of the standard normal distribution for $H(x)$ (①).

$$SPI = F^{-1}(H(x)) \tag{3}$$

To calculate the cumulative precipitation, which in turn determines drought, the precipitation when SPI is -1 can be calculated as Equation (4).

$$x = G^{-1}\left(\frac{F(-1) - q}{1 - q}\right) \tag{4}$$

where G^{-1} is the inverse CDF of the gamma distribution. To calculate the SPI for baseline and future climates, the non-exceedance probability of the gamma distribution for the 3-month cumulative precipitation during the future period was calculated, and the random variable of standard normal distribution with the same non-exceedance probability was obtained.

Figure 5 shows a method for calculating SPI and the meaning of change in the drought criteria by climate change. SPI is the random variable of the standard normal distribution corresponding to gamma CDF of specific precipitation. Even at the same precipitation, SPI varies according to change in the sample. In addition, criteria precipitation classifying the drought corresponding to SPI -1 also changes according to the sample. Therefore, magnitude and frequency of future drought can be changed depending on whether they are considering current (baseline) or future climate.

ⓐ and ⓑ in the figure represent criteria precipitation classifying drought. The left panel in Figure 5 shows CDF changes resulting in increasing precipitation of future compared to current (baseline). In this case, because the criteria precipitation corresponding to -1 increases, SPI values for the future is decreased (increased) on the basis of future or baseline (in case of employing gamma CDF of future or baseline) (Figure 5).

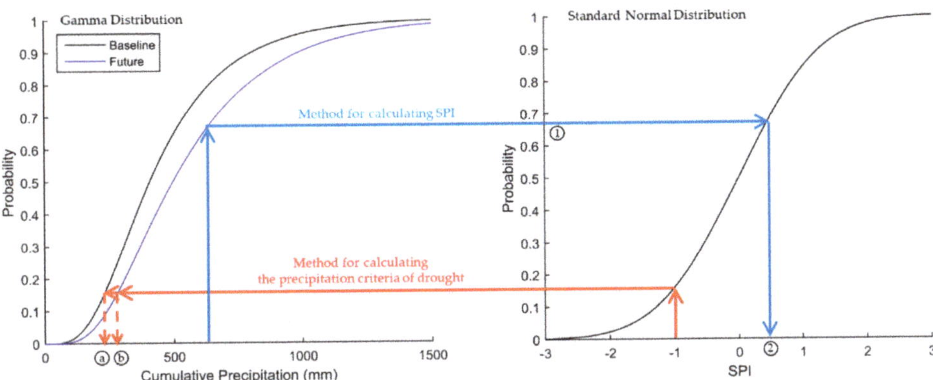

Figure 5. Concept of standardized precipitation index (SPI) and change in cumulative precipitation corresponding to SPI -1. As precipitation increases, the criteria of drought increase from (**a**) to (**b**).

3. Results

3.1. PDF Changes

This section analyzed the 3-month cumulative precipitation corresponding to SPI -1 during March, June, September, and December in the baseline and future climates. SPI values of ≤-1 define drought. The non-exceedance probability of the random variable of -1 is 0.1587, and the 3-month cumulative precipitation corresponding to 0.1587 was obtained using the inverse function of the gamma distribution. Figure 6 shows results for the mean cumulative precipitation of all stations in each period. Future 1 (2010–2039) is expected to decrease in March and September compared to the baseline climate, and increase in June and December. Although Future 2 (2040–2069) increased overall compared to Future 1, it is expected to decrease in June. Finally, Future 3 (2070–2099) decreased more in March and December than Future 2 did but increased in June and September (Figure 6 and Table 1).

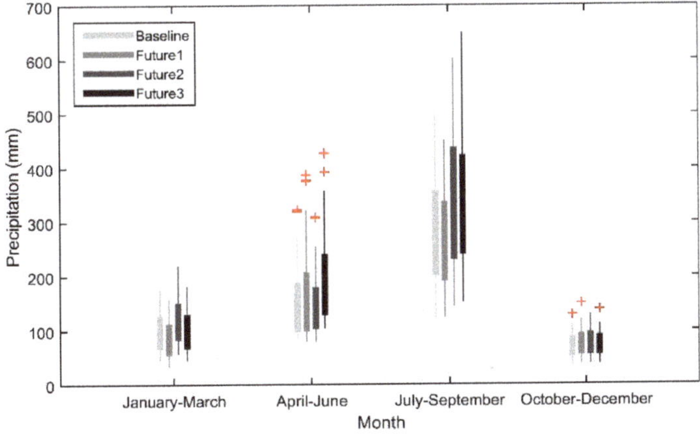

Figure 6. Box plot of 3-month cumulative precipitation corresponding to SPI of -1.

Table 1. Mean precipitation amounts (mm/3-months) corresponding to an SPI of -1.

Month	Baseline	Future		
		Future 1	Future 2	Future 3
March	97.3	86.8	120.1	100.7
June	152.7	169.0	150.6	197.8
September	270.3	257.3	326.1	328.8
December	70.3	75.6	78.1	72.7

The change of the gamma PDFs relative to the 3-month cumulative precipitation was analyzed for March, June, September, and December for the baseline and all future periods (Figure 7). Here, the band of each PDF indicates the range of the 95% confidence interval for all weather stations. Figure 7 shows that Future 1 is very similar to the baseline period, and the PDF shifts farther to the right between Future 2 and Future 3. Large precipitation amounts occur particularly frequently in Future 3, and the PDF of the mode is expected to be much smaller than for prior periods. Overall, 3-month cumulative precipitation increased in the future and the PDF shifted to the right. The average change in PDF was shown in Figure 7, it was projected that precipitation would increase in the future. However, due to the decrease in precipitation during autumn and winter, drought criteria precipitation (the precipitation which determines drought) was expected to decrease.

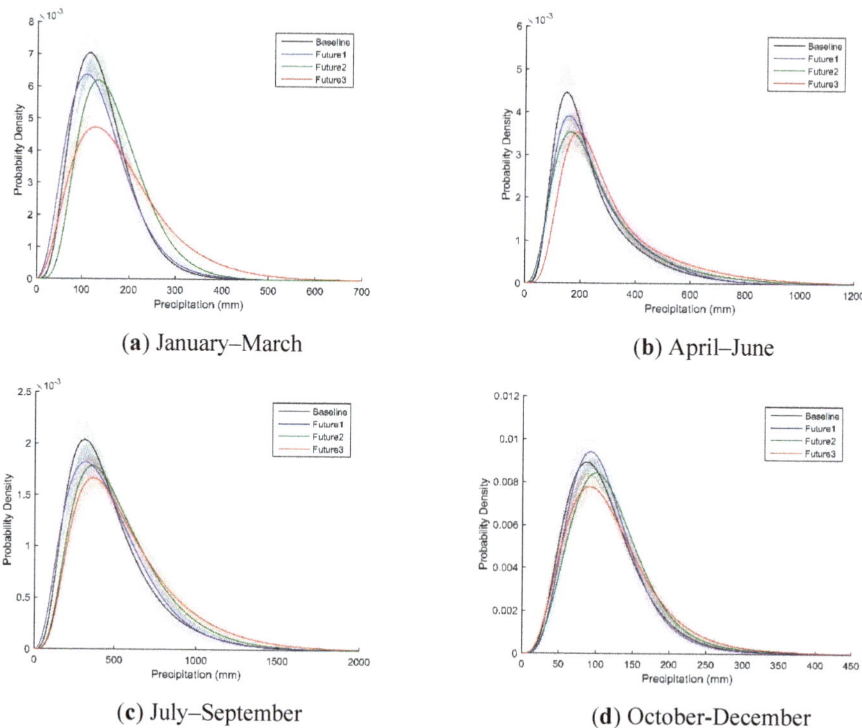

(**a**) January–March

(**b**) April–June

(**c**) July–September

(**d**) October-December

Figure 7. Probability density function (PDFs) of baseline and future 3-month precipitation.

Figure 8 shows the Coefficient of Variance (CV) for 3-month precipitation in baseline and future periods. The average CV values for the baseline and three future periods are 0.518, 0.517, 0.524, and 0.539, respectively; as such, CV was expected to increase in the future. Although larger CV means that the extreme climate is more likely to occur, it cannot be concluded that future drought is more severe. Therefore, we analyzed the future drought assessed in the baseline climate criteria in Section 3.2.

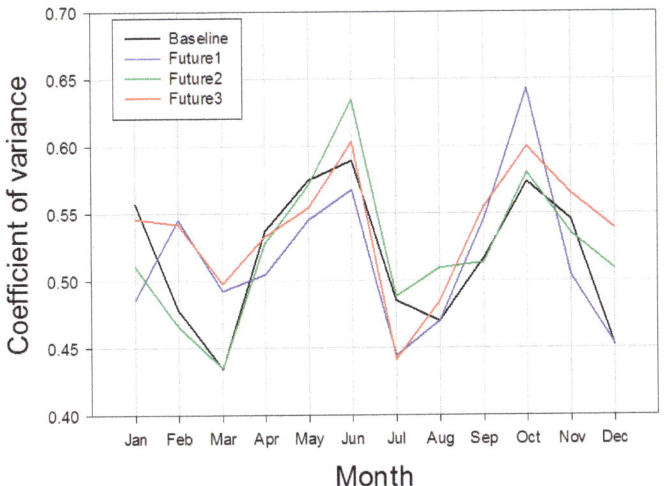

Figure 8. Coefficient of variance (CV) of baseline and future 3-month cumulative precipitation.

3.2. Changes in SPI

In Section 3.1, we identified changes in drought criteria precipitation for baseline and future climates. Results showed that drought criteria precipitation varied with each period due to changes in the PDFs, although overall precipitation increased over time. In this section, we compare a change in the drought index with the change in the reference period. To do so, the future SPI was calculated based on the baseline climate, and the severity and frequency of drought were compared (Figure 9).

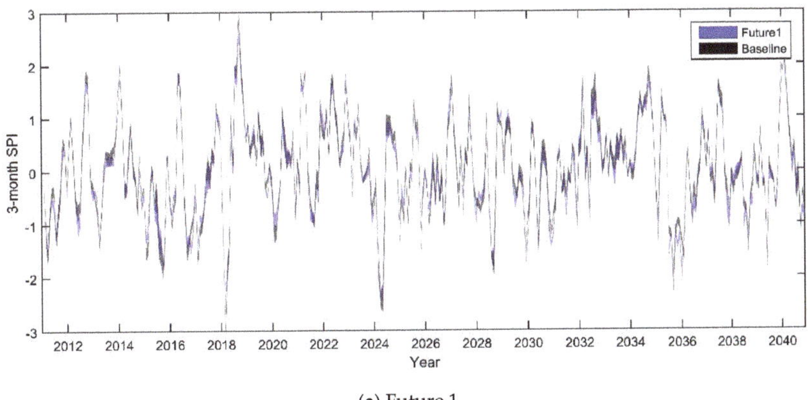

(**a**) Future 1

Figure 9. *Cont.*

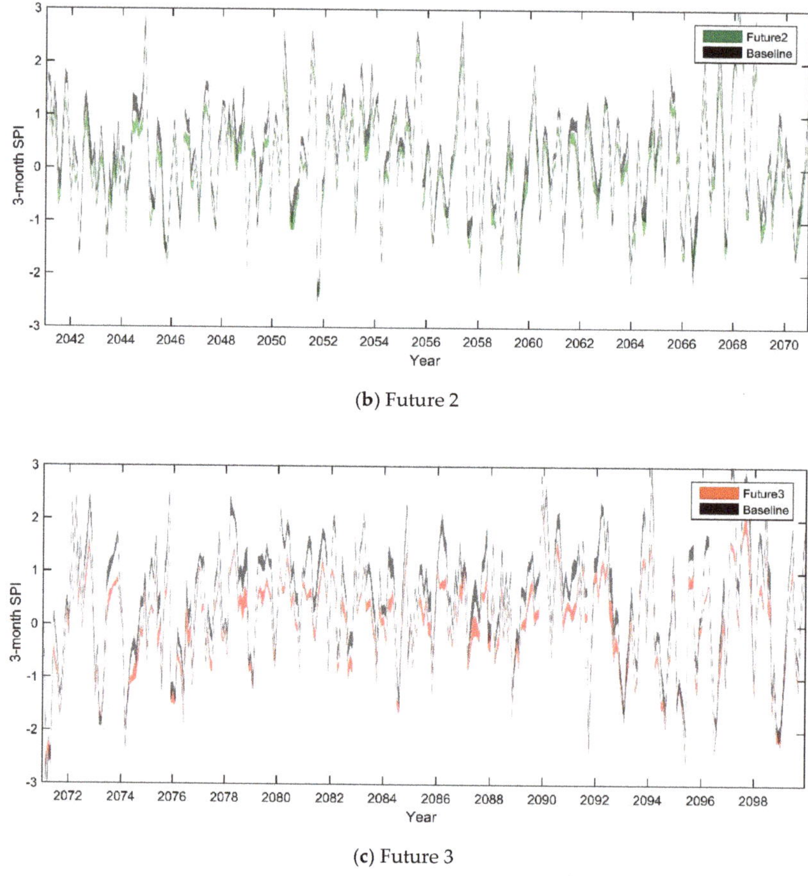

(**b**) Future 2

(**c**) Future 3

Figure 9. Future SPIs changes due to changes in the statistical characteristics of baseline and future climates.

Figure 9 shows SPIs for future precipitation under the baseline and future climate, with the 95% confidence interval shown for all weather stations. For the future precipitation, the SPI based on the future climate is smaller than the SPI based on the baseline climate. This is because of the increase in future precipitation within the climate change scenarios; therefore, the drought criteria precipitation was generally larger than the baseline. In addition, the smaller SPI is caused by an increasing occurrence frequency of less precipitation than the drought criteria precipitation, which happens because the variability of precipitation is increasing (Figure 9).

However, the difference between SPIs of the baseline and future climates increased as the reference period changed, and the SPI difference obtained from the baseline and Future 3 was the largest (Figure 9c). Overall, SPIs in the future climate were more severely evaluating drought. That is, under the baseline climate criterion, future cumulative precipitation did not represent drought conditions, yet it did indicate drought using the future climate criterion. Using the criterion of the baseline climate, the mean future SPIs were 0.04, 0.30, and 0.52, respectively. We can therefore conclude that future drought would be weaker under the baseline climate.

Table 2 shows the number of months that SPI were less than -1, -1.5, and -2.0 during various periods. Because the future climate has more precipitation than the baseline, drought frequency estimated by the baseline climate criterion is less than that estimated by the future climate criterion. In addition, moderate drought under the future climate criterion increased to 57.9, 57.4, and 59.4, respectively, but the frequency of drought under the baseline climate criterion decreased to 52.5, 42.5, and 41.4, respectively. From these results, we confirm that designs based on future climates may be exaggerated because of the higher frequency of drought under the future climate criterion than the baseline climate.

Table 2. Mean occurrence frequency of SPI less than or equal to −1.0, −1.5, and −2.0.

Criteria	Future 1		Future 2		Future 3	
	Baseline	Future 1	Baseline	Future 2	Baseline	Future 3
<−1.0	52.5	57.9	42.5	57.4	41.4	59.4
<−1.5	21.1	22.6	18.4	24.9	20.9	28.8
<−2.0	8.9	7.4	6.2	7.5	9.3	10.1

3.3. Spatial Distribution of Change in Drought Frequency

In Section 3.2, we compared drought indexes of future precipitation according to each parameter of gamma distribution; one is the parameters of baseline and others are parameters of future. On average, future drought frequency evaluated using the baseline climate criterion decreased. However, because the spatial distribution of drought occurrences vary widely due to regional climate, this section examines the spatial distribution of changes in drought frequency. For this, future drought frequency obtained from the baseline and future climate criteria was examined for each of the 57 weather stations (Figures 10–13). Figures 10–13 show the difference in the number of occurrences of future drought (SPI < −1), which is calculated as the number of drought occurrences using the future climate criterion minus that using the baseline climate criterion. In Figures 10–13, progressively redder colors indicate a larger number of droughts estimated using the future climate criterion.

Comparing the cumulative precipitation between baseline and future climate criteria for March of Future 2, a large region exhibited positive differences (Figure 10). In other words, in the Future 2 climate there are a lot of areas where drought will be more frequent than in the baseline climate. In most regions of South Korea, precipitation is projected to increase after the middle of the 21st century, relative to baseline values. As such, the drought criteria precipitation and drought frequency also increased. However, drought frequency using the Future 2 climate criterion is much larger than if the baseline climate criterion is used (Figure 10), showing the dependence on drought criteria precipitation.

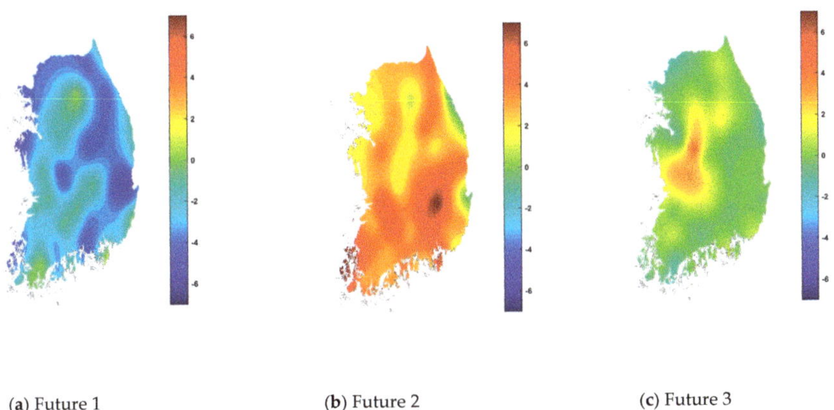

(**a**) Future 1   (**b**) Future 2   (**c**) Future 3

Figure 10. Future SPI changes in March due to changes in the statistical characteristics of the present and future climates.

In June, drought may be overestimated for all future periods, especially compared to March. In the case of Future 1, it was confirmed in Jeollanam-do and Chungcheongbuk-do, in Future

2, in Gyeonggi-do, and in Future 3, in Gyeonggi-do, Gangwon-do, and Gyeongsangnam-do. The difference was especially negative in the Gyeongsangnam-do of the Future 2 period. This is a region where frequent droughts occur relative to the baseline climate because the future precipitation in June is less than the baseline. This is because the gamma PDF of the future period is shifted to the left compared to the baseline due to less precipitation. As a result, these regions needed a strong drought response strategy to prepare for future droughts (Figure 11), which may be worse than projected.

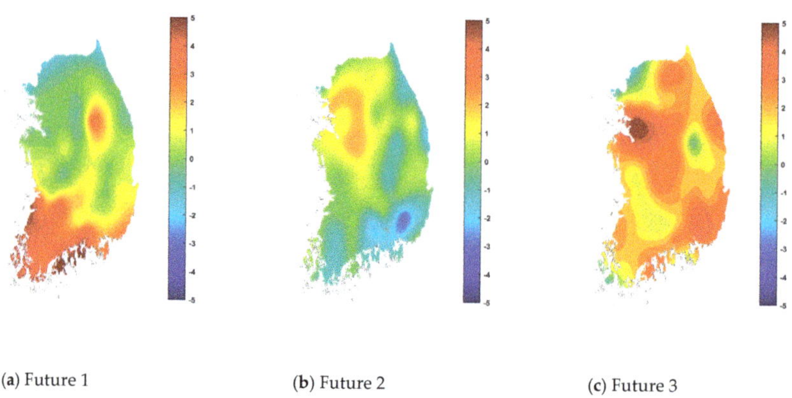

(**a**) Future 1   (**b**) Future 2   (**c**) Future 3

Figure 11. Future SPI changes in June due to changes in the statistical characteristics of the present and future climates.

In September in the Future 1 period, there was a broad area with a similar drought occurrence frequency between the baseline and future climate criteria, because there was no significant difference in drought criteria precipitation between the baseline and Future 1 climates. However, some regions such as Gangwon-do, exhibited negative differences. In Future 2, there was a wide region of positive differences due to the increase in criteria precipitation. On the west coast of Jeollanam-do and Jeollabuk-do, in some parts of Gangwon-do, Gyeongsangnam-do, and Gyeongsangbuk-do, positive difference regions were also observed in Future 3. Other areas are similar between future and baseline periods, and these regions are considered to be capable of coping with future droughts under a similar drought response strategy to that currently employed (Figure 12).

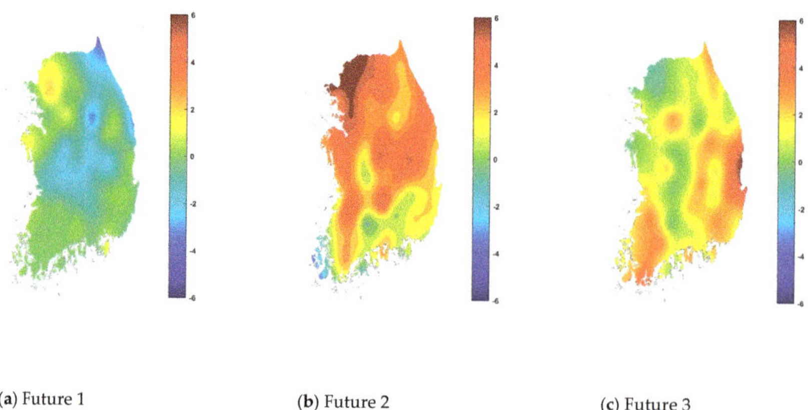

(**a**) Future 1   (**b**) Future 2   (**c**) Future 3

Figure 12. Future SPI changes in September due to changes in the statistical characteristics of the present and future climates.

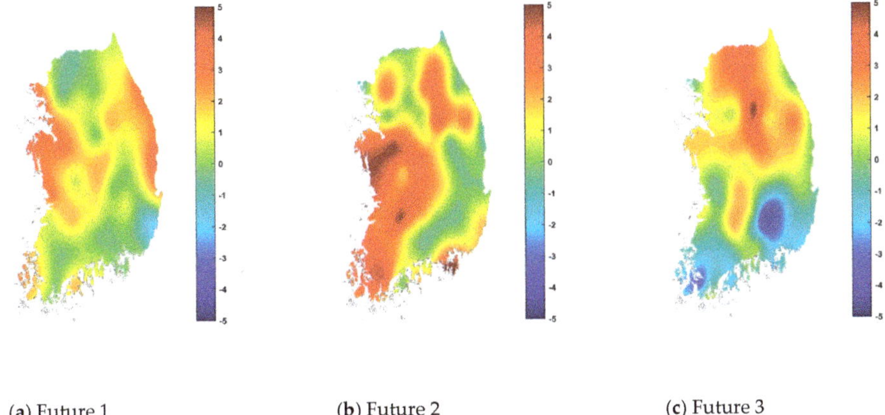

<table>
<tr><td>(a) Future 1</td><td>(b) Future 2</td><td>(c) Future 3</td></tr>
</table>

Figure 13. Future SPI changes in December due to changes in the statistical characteristics of the present and future climates.

In December, approximately half of South Korea experienced a larger drought frequency in Future 2 than the baseline period, due to the increase of precipitation in the future period. However, In Future 3, in most regions of Gyeongsangnam-do and Jeollanam-do, the opposite case was found. This is because future climate precipitation is less than the baseline, in most of the high-latitude regions in Future 3, drought has likely been overestimated due to the increase in future precipitation (Figure 13). Therefore, in these areas it may be sufficient to prepare for future drought using current mitigation strategies.

4. Conclusions

In general, studies that examined changes in future droughts calculated the frequency and severity of meteorological droughts as a criterion of a non-exceedance probability of precipitation over each period. Drought based on only future climates is difficult to quantitatively compare with the present since the drought criteria precipitation differs by period. Exceedance probability for the same quantitative value is required to prepare for extreme events, but drought studies to date have evaluated droughts based on the same non-exceedance probability for different criteria precipitation amounts. Therefore, this study predicted future drought changes based on the baseline climate.

Climate change scenarios were elucidated using statistical downscaling techniques representative of long-term climate change trends, and cumulative precipitation corresponding to drought criterion for each future period was calculated. Trends toward larger cumulative precipitation were generally confirmed, but some months in future period showed negative precipitation trends. Especially in September, when summer precipitation was included, variability was high. Moreover, results for the Future 1 period confirmed that it is important to prepare for drought due to lower expected precipitation compared to the baseline climate.

Kim et al. [15] projected that future droughts in South Korea will be more severe. They compared SPEI characteristics such as drought frequency and duration and found that mild drought frequency was projected to increase from the baseline 0.97/year to 3.72/year in the late of the 21st century, while severe drought was projected to increase from the baseline 0.20/year to 1.55/year in the late 21st century. Park et al. [21] predicted that drought would be severe in the Han River Basin in Korea under the RCP8.5 scenario. They indicated that drought severity will increase by approximately 45% in the Han River Basin as a result of the SDF (Severity-Duration-Frequency) curve. Kyoung et al. [20] projected that drought severity and frequency will increase in Seoul, while Park et al. [23] projected that drought duration and severity will increase in the central region of Korea, and Kim et al. [22] projected that drought frequency will increase in the Han River Basin. However, our results are different. As precipitation increases in the future, the PDF shifts to the right, and therefore the non-exceedance probability increases. Therefore, if future precipitation is assessed based on the baseline climate, our results showed that drought was weaker and less frequent. Therefore, the occurrence frequency of moderate drought based on the future climate tended to increase, but drought frequency based on baseline climate decreased. In other words, designing drought responses based on a future climate may be excessive.

However, future drought is underestimated in some sub-regions because the variation of the regional climate increases in the future. In particular, March in the Future 1 climate, June of Future 2, September of Future 1, and December of Future 3 exhibited less precipitation than the baseline climate, meaning that drought frequency was higher for the baseline climate criterion than the future climate criterion. Areas with these characteristics will require preparation for future drought based on the baseline climate. Our results suggest that regional priority can be assessed when constructing facilities for drought response.

Other studies for future drought have defined drought by the same non-exceedance probability for each period, rather than a comparison of quantities. This is similar to quantile mapping among the simplest downscaling method, and it is not suitable for a quantitative comparison between the baseline and future. It is significant that this study suggests and applies a method to evaluate future droughts based on the current climate. The inherent limitations to our study are that only one GCM model and RCP were used. Especially, when there was a difference between HadGEM2-AO and observation during the rainy seasons, from this study, future water availability can be assessed realistically, and climate change scenario uncertainty can be quantified in the future by using additional models and RCPs.

Author Contributions: Conceptualization, J.H.S.; formal analysis, J.H.S. and M.K.; investigation, J.H.S. and M.K.; data curation, J.H.S.; writing—original draft preparation, J.H.S.; writing—review and editing, J.H.S. and M.K.

Funding: Funding: This research was funded by the Korea Environmental Industry & Technology Institute (KEITI) grant funded by the Ministry of Environment grant number Grant 18AWMP-B083066-05.

References

1. Wilhite, D. Drought monitoring and early warning: Concepts, progress and future challenges. *World Meteorol. Organ.* **2006**, *1006*, 24.

2. WMO. Experts agree on a universal drought index to cope with climate risks. WMO Press Release 2009, No. 872. Available online: http://www.wmo.int/pages/prog/wcp/agm/meetings/wies09/documents/872_en.pdf (accessed on 17 July 2018).

3. Stokes, B.; Wike, R.; Carle, J. Global concern about climate change, broad support for limiting emissions. 2015. Available online: http://www.pewglobal.org/2015/11/05/global-concern-about-climate-change-broad-support-for-limiting-emissions/ (accessed on 17 July 2018).

4. Wilhite, D.A.; Glantz, M.H. Understanding: The drought phenomenon: The role of definitions. *Water Int.* **1985**, *10*, 111–120. [CrossRef]

5. American Meteorological Society. Meteorological drought—Policy statement. *Bull. Am. Meteorol. Soc.* **1997**, *78*, 847–849. [CrossRef]

6. Heim, R.R., Jr. A review of twentieth-century drought indices used in the United States. *Bull. Am. Meteorol. Soc.* **2002**, *83*, 1149–1165. [CrossRef]

7. McKee, T.B.; Doesken, N.J.; Kleist, J. The relationship of drought frequency and duration to time scales. In *Proceedings of the 8th Conference on Applied Climatology*; American Meteorological Society: Boston, MA, USA, 1993; pp. 179–183.

8. Zhai, J.; Su, B.; Krysanova, V.; Vetter, T.; Gao, C.; Jiang, T. Spatial variation and trends in PDSI and SPI indices and their relation to streamflow in 10 large regions of china. *J. Clim.* **2010**, *23*, 649–663. [CrossRef]

9. Lorenzo-Lacruz, J.; Vicente-Serrano, S.M.; López-Moreno, J.I.; Beguería, S.; García-Ruiz, J.M.; Cuadrat, J.M. The impact of droughts and water management on various hydrological systems in the headwaters of the Tagus River (central Spain). *J. Hydrol.* **2010**, *386*, 13–26. [CrossRef]

10. Eum, H.I.; Cannon, A.J. Intercomparison of projected changes in climate extremes for South Korea: application of trend preserving statistical downscaling methods to the CMIP5 ensemble. *Int. J. Climatol.* **2017**, *37*, 3381–3397. [CrossRef]

11. Sung, J.H.; Kang, H.-S.; Park, S.; Cho, C.; Bae, D.H.; Kim, Y.-O. Projection of extreme precipitation at the end of 21st century over South Korea based on representative concentration pathways (RCP). *Atmosphere* **2012**, *22*, 221–231. [CrossRef]

12. Sung, J.H.; Kim, B.S.; Kang, H.-S.; Cho, C. Nonstationary frequency analysis for extreme precipitation based on Representative Concentration Pathways (RCP) climate change scenarios. *J. Korean Soc. Hazard Mitig.* **2012**, *12*, 231–244. [CrossRef]

13. Suh, M.S.; Oh, S.G.; Lee, D.K.; Cha, D.H.; Choi, S.J.; Jin, C.S.; Hong, S.Y. Development of new ensemble methods based on the performance skills of regional climate models over South Korea. *J. Clim.* **2012**, *25*, 7067–7082. [CrossRef]

14. Kwon, M.; Lee, G.; Jun, K.S. Analysis of Annual Maximum Daily Rainfall Using RCP Climate Change Scenario in Korean Peninsula. *J. Korean Soc. Hazard Mitig.* **2015**, *15*, 99–110. [CrossRef]

15. Kim, B.S.; Chang, I.G.; Sung, J.H.; Han, H.J. Projection in Future Drought Hazard of South Korea Based on RCP Climate Change Scenario 8.5 Using SPEI. *Adv. Meteorol.* **2016**. Available online: https://www.hindawi.com/journals/amete/2016/4148710/abs/ (accessed on 11 February 2019).

16. Riahi, K.; Rao, S.; Krey, V.; Cho, C.; Chirkov, V.; Fischer, G.; Rafaj, P. RCP 8.5—A scenario of comparatively high greenhouse gas emissions. *Clim. Chang.* **2011**, *109*, 33. [CrossRef]

17. Vicente-Serrano, S.M.; Beguería, S.; López-Moreno, J.I. A multiscalar drought index sensitive to global warming: the standardized precipitation evapotranspiration index. *J. Clim.* **2010**, *23*, 1696–1718. [CrossRef]

18. Park, J.; Lim, Y.-J.; Kim, B.-J.; Sung, J.H. Appraisal of Drought Characteristics of Representative Drought Indices using Meteorological Variables. *KSCE J. Civ. Eng.* **2018**, *22*, 2002–2009. [CrossRef]

19. Lee, J.-H.; Kwon, H.-H.; Jang, H.-W.; Kim, T.-W. Future changes in drought characteristics under extreme climate change over South Korea. *Adv. Meteorol.* **2016**. Available online: https://www.hindawi.com/journals/amete/2016/9164265/abs/ (accessed on 11 February 2019). [CrossRef]

20. Kyoung, M.; Lee, Y.; Kim, H.; Kim, B. Assessment of climate change effect on temperature and drought in Seoul: Based on the AR4 SRES A2 senario. *J. Korean Soc. Civ. Eng.* **2009**, *29*, 181–191.

21. Park, B.-S.; Lee, J.-H.; Kim, C.-J.; Jang, H.-W. Projection of future drought of Korea based on probabilistic approach using multi-model and multi climate change scenarios. *J. Korean Soc. Civ. Eng.* **2013**, *33*, 1871–1885. [CrossRef]

22. Kim, D.; Hong, S.J.; Han, D.; Choi, C.; Kim, H.S. Analysis of Future Meteorological Drought Index Considering Climate Change in Han-River Basin. *J. Wes. Res.* **2016**, *18*, 432–447.

23. Park, M.; Lee, O.; Park, Y.; Kim, S. Future drought projection In Korea under ar5 rcp climate change scenarios. *J. Korean Soc. Hazard Mitig.* **2015**, *15*, 423–433. [CrossRef]

24. Collins, W.J.; Bellouin, N.; Doutriaux-Boucher, M.; Gedney, N.; Halloran, P.; Hinton, T.; Hughes, J.; Jones, C.D.; Joshi, M.; Lid-dicoat, S.; et al. Development and evaluation of an Earth-system model – HadGEM2. *Geosci. Model Dev.* **2011**, *4*, 997–1062. [CrossRef]

25. IPCC. Climate change 2014: Synthesis report. In *Contribution of Working Groups I. II and III to the Fifth Assessment Report of the Intergovernmental Panel on Climate Change*; IPCC: Geneva, Switzerland, 2014.

26. Cho, J.; Jung, I.; Cho, W.; Hwang, S. User-centered climate change scenarios technique development and application of Korean peninsula. *J. Clim. Chang. Res.* **2018**, *9*, 13–29. [CrossRef]

27. Cannon, A.J.; Sobie, S.R.; Murdock, T.Q. Bias correction of GCM precipitation by quantile mapping: How well do methods preserve changes in quantiles and extremes? *J. Clim.* **2015**, *28*, 6938–6959. [CrossRef]

28. Bürger, G.; Sobie, S.R.; Cannon, A.J.; Werner, A.T.; Murdock, T.Q. Downscaling extremes: An intercomparison of multiple methods for future climate. *J. Clim.* **2013**, *26*, 3429–3449. [CrossRef]

29. Hong, J.Y.; Ahn, J.B. Changes of early summer precipitation in the Korean Peninsula and nearby regions based on RCP simulations. *J. Clim.* **2015**, *28*, 3557–3578. [CrossRef]

30. Lee, H.S.; Gan, S.Y.; Byun, Y.H.; Kang, H.S.; Hyun, Y.K.; Haek, H.J.; Kwon, W.T. *Evaluation of HadGEM2-AO based on historical simulation of IPCC AR5*; Korean Meteorological Society: Daegu, South Korea, 2009.
31. Lee, H.S.; Gan, S.Y.; Baek, H.J.; Cho, C.H. *Evaluation of the pre-industrial simulation of HadGEM2-AO*; Korean Meteorological Society: Busan, Korea, 2010.

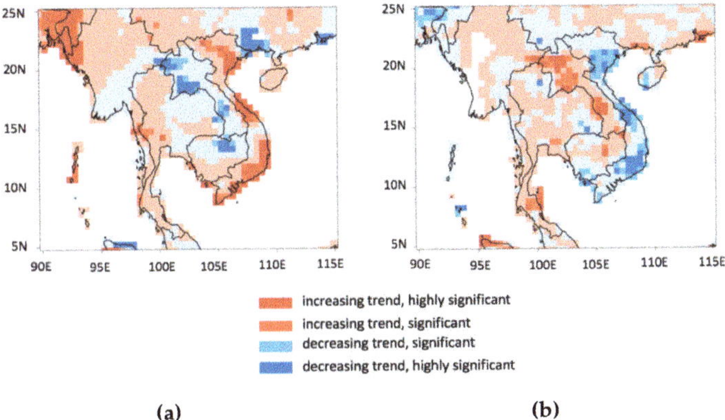

(**a**)   (**b**)

Figure 2. Trends in (**a**) spring precipitation and (**b**) extreme drought in the Indochina Peninsula. The modified Mann–Kendall (M–K) test was applied; statistically significant changes are highlighted in red (increasing trend) and blue (decreasing trend) for 90% (significant) and 95% (highly significant) confidence intervals.

3.2. Composite Anomaly of Extreme Drought for IOD Modes

Figure 3 shows the composite anomalies of extreme drought during the spring season (March–May) for positive and negative IOD years. During the positive IOD years, negative anomaly patterns of drought were dominant throughout Vietnam and the southwestern region of the Yunnan Province in China. However, a positive anomaly became evident over central Cambodia and Laos, with distinct abnormalities occurring along the coastline adjacent to the Myanmar Sea; in some areas, abnormalities recorded were more than 160% higher than usual. Furthermore, if both positive IOD years and El Niño events occurred simultaneously, the spatial distribution of extreme drought was observed to intensify only compared to drought anomaly patterns during the positive IOD years. The change in extreme drought during negative IOD years was similar to the change in positive IOD years. Positive anomaly of extreme drought around Myanmar's coastline was weakened, and positive anomalies were more evident across Cambodia. The results for negative IOD years were similar to the changes for positive IOD years; however, the anomalies increased towards Northern Laos, Northeastern Myanmar, and Northern Thailand. These changes were more evident when the negative IOD years coincided with La Niña events.

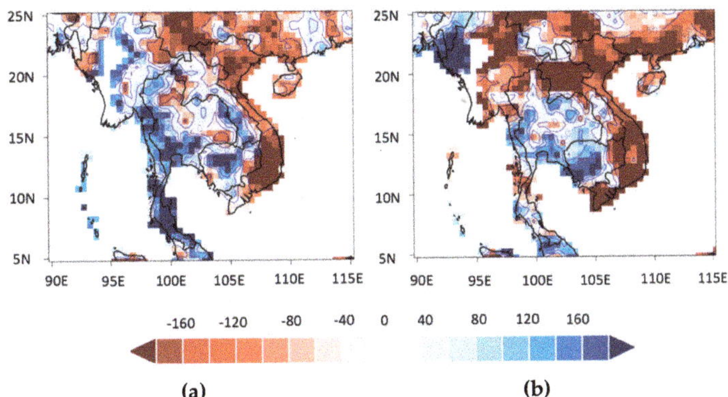

(**a**)   (**b**)

Figure 3. Percentage composite anomalies in extreme drought (departures from the 1981–2010 norms) during (**a**) positive IOD years and (**b**) negative IOD years.

3.3. Changes in Extreme Drought Using the IBB method

Figures 4 and 5 show the results of composite anomalies of extreme drought (shown as a percentage of the normal) during spring in relation to positive and negative IOD simulations over the ICP. An increase of 0.25σ in IOD (Figure 4) resulted in more extreme drought across Laos and along the coastline adjacent to the Myanmar Sea, and less extreme droughts than usual across Southern Cambodia, Vietnam, and Northwest Myanmar, adjacent to the Bay of Bengal. As IOD increases, drought changes appear more severe. The northwestern region of Myanmar was observed to have a statistically significant increase in extreme drought for more than 0.5σ of the IOD. Some parts of Southern Thailand and Western Cambodia showed a statistically significant negative anomaly pattern of extreme drought. These patterns tended to be reinforced when drought symptoms appeared from +0.75σ to +1σ of the IOD. In addition, statistically significant patterns of positive anomalies in extreme droughts in some areas along the coastline of the Myanmar Sea were observed in cases of more than +1σ of IOD changes.

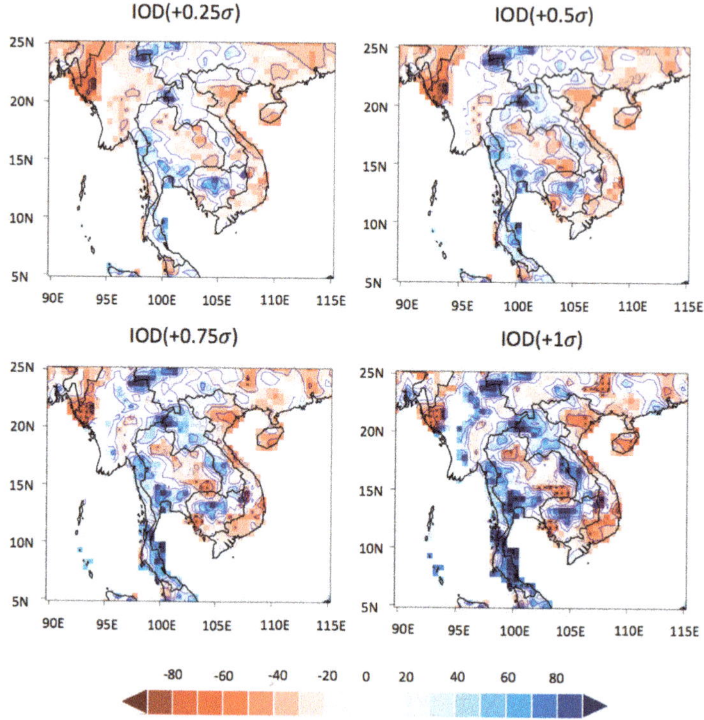

Figure 4. Percentage changes in extreme drought over the Indochina Peninsula (departures from the 1981–2010 norms). The climatology mean of IOD (1981–2010) was deliberately increased by 0.25σ through IBB simulations to analyze the sensitivity of extreme drought to the warming phase of IOD in the tropical Indian Ocean. In each figure, the median results derived from 1000 simulations are shown and the statistically significant area of change at 95% significance level is represented by "x".

Figure 5 shows a 0.25σ decrease in the IOD. When negative IODs were strengthened, it was found that extreme droughts were less frequent throughout Laos, Vietnam, and the southwestern region of the Yunnan Province in China. However, in Northwestern Myanmar, some parts of Thailand, and in regions adjacent to the Andaman Sea, extreme drought increased, leading to greater drought vulnerability. In addition, IOD-sensitive areas with more extreme droughts emerged in the northwestern parts of

Myanmar, especially as changes in negative IOD became apparent. However, the eastern and northern parts of the ICP showed a contrasting response to negative IOD. The simulated results indicated that extreme droughts throughout Northwestern Myanmar and Thailand were more frequent than usual.

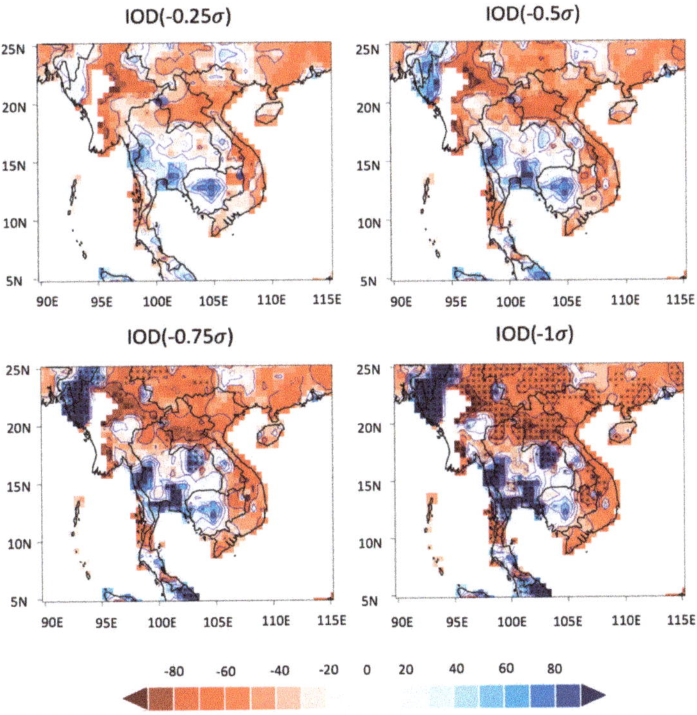

Figure 5. As Figure 4, but for a decrease in IOD in the tropical Indian Ocean.

4. Discussion and Conclusions

IOD in the Indian Ocean and ENSO in the Pacific Ocean are the results of atmospheric–ocean interactions in each region. Both phenomena can be identified by SST anomalies and are reported to have a direct impact on rainfall occurring globally [14,17,34–36]. The ENSO phenomenon has been officially observed since 1985 [37] and is relatively well known for its anomalous convection, wind patterns, and predictable short–term behaviors [38–41] within the global climate system. However, there is still a lack of understanding of interactions, evolutionary patterns, and predictability of new forms of El Niño [42].

Long-term drought conditions in Southeast Asia generally relate to El Niño, while wetter conditions tend to be associated with La Nina [43]. These changes are consistent with the results of the composite analysis of IOD and ENSO conducted in this study. When positive IOD and El Niño occur simultaneously, the spatial distribution of spring rainfall is significantly reduced compared to the rainfall anomaly pattern that takes into account positive IOD years only. Results for negative IOD years are similar to the changes observed for positive IOD years, but it is expanding to Northern Laos, Northeastern Myanmar, and Northern Thailand. These changes are more evident when the negative IOD years coincide with La Niña events.

Although different forms of El Niño have been reported to occur with increasing frequency in the central Pacific since the 2000s [36,40], the exact causes of these occurrences and their impacts on the Indian Ocean and Pacific Rim countries—according to the evolution patterns of different types of El Niño events—are still unknown. Identification of different types of El Niño is evidence that ENSO events are not well understood and are certainly unpredictable [42]. Furthermore, ENSO and IOD–rainfall linkages in the ICP show significant spatial and seasonal variations. Recent studies suggested that the slowdown in the observed global average atmospheric temperature rise is closely related to substantial heat transfer from the Pacific Ocean to the Indian Ocean through Indonesian throughflow [44–46]. Despite similarities with ENSO, the IOD is an even more recent discovery with even less understood implications [14,18].

The spatio-temporal connection between SST and winds shows a strong coupling of precipitation and ocean dynamics in the Indian Ocean. However, with regards to the IOD phenomena, few quantitative studies provide a complete picture of extreme drought fluctuations across the ICP. Therefore, in this study, areas most vulnerable to drought were analyzed through statistical simulations of IODs using the IBB method based on historical observations.

The results of the long-term trend analysis show that the areas where spring rainfall has increased are mainly distributed along the eastern coast (Vietnam) and the northwestern portions of the ICP, while central and Northern Laos and Northern Cambodia have experienced a reduction in spring rainfall over the past few decades. This is because of the Truong-Son Mountains in Vietnam, which are adjacent to the South China Sea, have relatively abundant rainfall, whereas Laos has seen relatively little rainfall. This trend is similar to that of extreme droughts. Vietnam experienced less severe droughts, but central and Northern Laos suffered more severe droughts. During positive IOD years, extreme drought occurred less than usual throughout Vietnam and Southwestern China, while more droughts were observed in Cambodia and central Laos, as well as along the coastline adjacent to the Myanmar Sea. Results for negative IOD years are similar to changes observed for positive IOD years; however, the eastern and northern parts of the ICP are experiencing a decrease in droughts. In addition, the results of the statistical simulations proposed in this study successfully simulate drought-sensitive areas and evolution patterns for various IOD changes.

Investigating SST changes in the Indo-Pacific Ocean can help improve the understanding of regional–scale climatic variability as well as quantitatively diagnose this variability. Moreover, the sensitivity analysis of extreme drought to IODs based on the statistical simulations demonstrates the importance of further investigation of the mechanisms of IOD impacts on drought evolution in the ICP, which may include complex coupled oceanic–atmospheric processes. Although the results obtained herein were based on limited observations, the study's findings have strong implications for improving extreme drought predictions in the ICP.

Author Contributions: conceptualization, J.-S.K.; Formal analysis, Q.-G.G. and J.-S.K.; Methodology, J.-H.L., V.S., and J.-S.K.; resources, V.S.; writing—original draft preparation, Q.-G.G. and J.-S.K.; writing—review and editing, L.X., J.-S.K., and J.-H.L.

Funding: This research is supported by the National Key R&D Program of China (2017YFC0405901), the National Natural Science Foundation of China (No. 51525902), and the Ministry of Education "111 Project" Fund of China (B18037), all of which are greatly appreciated. The fourth author was supported by Basic Science Research Program through the National Research Foundation of Korea (NRF) funded by the Ministry of Education (NRF-2017R1D1A1A02018546). The APC was funded by the National Key R&D Program of China (2017YFC0405901).

Conflicts of Interest: The authors declare no conflicts of interest.

References

1. Dai, A. Drought under global warming: A review. *Wiley Interdiscip. Rev. Clim. Change* **2011**, *2*, 45–65. [CrossRef]

2. Dai, A. Increasing drought under global warming in observations and models. *Nat. Clim. Change* **2013**, *3*, 52–58. [CrossRef]

3. Rind, D.; Goldberg, R.; Hansen, J.; Rosenzweig, C.; Ruedy, R. Potential evapotranspiration and the likelihood of future drought. *J. Geophys. Res.: Atmos.* **1990**, *95*, 9983–10004. [CrossRef]

4. Wang, G. Agricultural drought in a future climate: results from 15 global climate models participating in the IPCC 4th assessment. *Clim. Dyn.* **2005**, *25*, 739–753. [CrossRef]

5. Richard, R.; Heim, J. A Review of Twentieth-Century Drought Indices Used in the United States. *Bull. Am. Meteorol. Soc.* **2002**, *83*, 1149–1166.

6. Gao, Q.; Kim, J.-S.; Chen, J.; Chen, H.; Lee, J.-H. Atmospheric Teleconnection-Based Extreme Drought Prediction in the Core Drought Region in China. *Water* **2019**, *11*, 232. [CrossRef]

7. Le, M.H.; Perez, G.C.; Solomatine, D. Meteorological Drought Forecasting Based on Climate Signals Using Artificial Neural Network — A Case Study in Khanhhoa Province Vietnam. *Procedia Eng.* **2016**, *154*, 1169–1175. [CrossRef]

8. Nguyen, H.; Shaw, R. Chapter 8 Drought Risk Management in Vietnam. In *Droughts in Asian Monsoon Region*; Emerald Group Publishing Limited: Bingley, UK, 2011; pp. 141–161.

9. Hundertmark, W.J.I. Building drought management capacity in the Mekong river basin. *Irrig. Drain.* **2008**, *57*, 279–287. [CrossRef]

10. Vu, M.T.; Raghavan, V.S.; Liong, S.-Y. Ensemble Climate Projection for Hydro-Meteorological Drought over a river basin in Central Highland, Vietnam. *KSCE J. Civ. Eng.* **2015**, *19*, 427–433. [CrossRef]

11. Shu, F.; Prasoot, S.; Monthathip, C. Increasing Production of Rainfed Lowland Rice in Drought Prone Environments. *Plant Prod. Sci.* **1998**, *1*, 75–82.

12. Kim, J.-S.; Seo, G.-S.; Jang, H.-W.; Lee, J.-H. Correlation analysis between Korean spring drought and large-scale teleconnection patterns for drought forecasting. *KSCE J. Civ. Eng.* **2017**, *21*, 458–466. [CrossRef]

13. Zhang, Z.; Jin, Q.; Chen, X.; Xu, C.-Y.; Jiang, S.J. On the Linkage between the Extreme Drought and Pluvial Patterns in China and the Large-Scale Atmospheric Circulation. *Adv. Meteorol.* **2016**, *2016*. [CrossRef]

14. Saji, N.H.; Goswami, B.N.; Vinayachandran, P.N.; Yamagata, T. A dipole mode in the tropical Indian Ocean. *Nature* **1999**, *401*, 360–363. [CrossRef] [PubMed]

15. Jin, F.-F. An Equatorial Ocean Recharge Paradigm for ENSO. Part I: Conceptual Model. *J. Atmos. Sci.* **1997**, *54*, 811–829. [CrossRef]

16. Black, E.; Slingo, J.; Sperber, K.R. An Observational Study of the Relationship between Excessively Strong Short Rains in Coastal East Africa and Indian Ocean SST. *Mon. Weather Rev.* **2003**, *131*, 74–94. [CrossRef]

17. Ashok, K.; Guan, Z.; Yamagata, T. Impact of the Indian Ocean Dipole on the Relationship between the Indian Monsoon Rainfall and ENSO. *Geophys. Res. Lett.* **2001**, *28*, 4499–4502. [CrossRef]

18. Saji, N.; Yamagata, T. Possible impacts of Indian Ocean Dipole mode events on global climate. *Clim. Res.* **2003**, *25*, 151–169. [CrossRef]

19. Ashok, K.; Guan, Z.; Yamagata, T. Influence of the Indian Ocean Dipole on the Australian winter rainfall. *Geophys. Res. Lett.* **2003**, *30*, 1821. [CrossRef]

20. Liess, S.; Kumar, A.; Snyder, P.K.; Kawale, J.; Steinhaeuser, K.; Semazzi, F.H.M.; Ganguly, A.R.; Samatova, N.F.; Kumar, V. Different Modes of Variability over the Tasman Sea: Implications for Regional Climate. *J. Clim.* **2014**, *27*, 8466–8486. [CrossRef]

21. Cai, W.; Cowan, T.; Sullivan, A. Recent unprecedented skewness towards positive Indian Ocean Dipole occurrences and its impact on Australian rainfall. *Geophys. Res. Lett.* **2009**, *36*, 245–253. [CrossRef]

22. Chan, S.C.; Behera, S.K.; Yamagata, T. Indian Ocean Dipole influence on South American rainfall. *Geophys. Res. Lett.* **2008**, *35*, L14S12. [CrossRef]

23. Zanchettin, D.; Bothe, O.; Müller, W.; Bader, J.; Jungclaus, J.H. Different flavors of the Atlantic Multidecadal Variability. *Clim. Dyn.* **2014**, *42*, 381–399. [CrossRef]

24. Cai, W.; Zheng, X.-T.; Weller, E.; Collins, M.; Cowan, T.; Lengaigne, M.; Yu, W.; Yamagata, T. Projected response of the Indian Ocean Dipole to greenhouse warming. *Nat. Geosci.* **2013**, *6*, 999–1007. [CrossRef]

25. Ihara, C.; Kushnir, Y.; Cane, M.A. Warming Trend of the Indian Ocean SST and Indian Ocean Dipole from 1880 to 2004. *J. Clim.* **2008**, *21*, 2035–2046. [CrossRef]

26. Ren, H.-L.; Jin, F.-F. Niño indices for two types of ENSO. *Geophys. Res. Lett.* **2011**, *38*. [CrossRef]

27. McKee, T.B.; Doesken, N.J.; Kleist, J. The relationship of drought frequency and duration to time scales. In Proceedings of the 8th Conference on Applied Climatology, 8th Conference on Applied Climatology, Anaheim, CA, USA, 17–22 January 1993; American Meteorological Society: Boston, MA, USA, 1993; pp. 179–183.

28. Guttman, N.B. ACCEPTING THE STANDARDIZED PRECIPITATION INDEX: A CALCULATION ALGORITHM. *J. Am. Water Resour. Assoc.* **1999**, *35*, 311–322. [CrossRef]

29. Mann, H.B. Nonparametric Tests Against Trend. *Econometrica* **1945**, *13*, 245–259. [CrossRef]
30. Kendall, M.G. *Rank correlation methods*; Griffin: Oxford, UK, 1948.
31. Hamed, K.H.; Rao, A.R. A modified Mann-Kendall trend test for autocorrelated data. *J. Hydrol.* **1998**, *204*, 182–196. [CrossRef]
32. Hall, P.; Presnell, B. Intentionally biased bootstrap methods. *J. R. Stat. Soc.* **1999**, *61*, 143–158. [CrossRef]
33. Lee, T. Climate change inspector with intentionally biased bootstrapping (CCIIBB ver. 1.0) — methodology development. *Geosci. Model Dev.* **2017**, *10*, 525–536. [CrossRef]
34. Piechota, T.C.; Chiew, F.H.S.; Dracup, J.A.; McMahon, T.A. Seasonal streamflow forecasting in eastern Australia and the El Niño–Southern Oscillation. *Water Resour. Res.* **1998**, *34*, 3035–3044. [CrossRef]
35. Mahala, B.K.; Nayak, B.K.; Mohanty, P.K. Impacts of ENSO and IOD on tropical cyclone activity in the Bay of Bengal. *Nat. Hazards* **2015**, *75*, 1105–1125. [CrossRef]
36. Pradhan, P.K.; Preethi, B.; Ashok, K.; Krishna, R.; Sahai, A.K. Modoki, Indian Ocean Dipole, and western North Pacific typhoons: Possible implications for extreme events. *J. Geophys. Res.* **2011**, *116*, D18108. [CrossRef]
37. McPhaden, M.J.; Ando, K.; Bourles, B.; Freitag, H.; Lumpkin, R.; Masumoto, Y.; Murty, V.; Nobre, P.; Ravichandran, M.; Vialard, J.J.P.O.O. The Global Tropical Moored Buoy Array. In Proceedings of the "OceanObs'09: Sustained Ocean Observations and Information for Society" Conference, OceanObs'09, Venice, Italy, 21–25 September 2009.
38. Kim, J.-S.; Son, C.-Y.; Moon, Y.-I.; Lee, J.-H. Seasonal rainfall variability in Korea within the context of different evolution patterns of the central Pacific El Niño. *J. Water Clim. Change* **2017**, *8*, 412–422. [CrossRef]
39. Kim, J.-S.; Kim, S.T.; Wang, L.; Wang, X.; Moon, Y.-I. Tropical cyclone activity in the northwestern Pacific associated with decaying Central Pacific El Niños. *Stochastic Environ. Res. Risk Assess.* **2016**, *30*, 1335–1345. [CrossRef]
40. Kim, J.-S.; Zhou, W.; Wang, X.; Jain, S. El Niño Modoki and the Summer Precipitation Variability over South Korea: A Diagnostic Study. *J. Meteorolog. Soc. Jpn.* **2012**, *90*, 673–684. [CrossRef]
41. Ashok, K.; Yamagata, T. Climate change: The El Niño with a difference. *Nature* **2009**, *461*, 481–484. [CrossRef] [PubMed]
42. Chiodi, A.M.; Harrison, D.E. Observed El Niño SSTA Development and the Effects of Easterly and Westerly Wind Events in 2014/15. *J. Clim.* **2017**, *30*, 1505–1519. [CrossRef]
43. Juneng, L.; Tangang, F.T. Evolution of ENSO-related rainfall anomalies in Southeast Asia region and its relationship with atmosphere–ocean variations in Indo-Pacific sector. *Clim. Dyn.* **2005**, *25*, 337–350. [CrossRef]
44. Kosaka, Y.; Xie, S.-P. Recent global-warming hiatus tied to equatorial Pacific surface cooling. *Nature* **2013**, *501*, 403. [CrossRef]
45. Lee, S.-K.; Park, W.; Baringer, M.O.; Gordon, A.L.; Huber, B.; Liu, Y. Pacific origin of the abrupt increase in Indian Ocean heat content during the warming hiatus. *Nat. Geosci.* **2015**, *8*, 445–449. [CrossRef]
46. Liu, W.; Xie, S.-P.; Lu, J. Tracking ocean heat uptake during the surface warming hiatus. *Nat. Commun.* **2016**, *7*, 10926. [CrossRef] [PubMed]

Article

Atmospheric Teleconnection-Based Extreme Drought Prediction in the Core Drought Region in China

Qinggang Gao [1], Jong-Suk Kim [1,*], Jie Chen [1], Hua Chen [1] and Joo-Heon Lee [2,*]

[1] State Key Laboratory of Water Resources and Hydropower Engineering Science, Wuhan University, Wuhan 430072, China; 2014301580312@whu.edu.cn (Q.G.); jiechen@whu.edu.cn (J.C.); chua@whu.edu.cn (H.C.)

[2] Department of Civil Engineering, Joongbu University, Go Yang, Gyeong Gi 10279, Korea

* Correspondence: jongsuk@whu.edu.cn (J.-S.K.); leejh@joongbu.ac.kr (J.-H.L.)

Received: 29 December 2018; Accepted: 27 January 2019; Published: 30 January 2019

Abstract: This paper aims to improve the predictability of extreme droughts in China by identifying their relationship with atmospheric teleconnection patterns (ATPs). Firstly, a core drought region (CDR) is defined based on historical drought analysis to investigate possible prediction methods. Through the investigation of the spatial-temporal characteristics of spring drought using a modified Mann–Kendall test, the CDR is found to be under a decadal drying trend. Using principal component analysis, four principal components (PCs), which explain 97% of the total variance, are chosen out of eight teleconnection indices. The tree-based model reveals that PC1 and PC2 can be divided into three groups, in which extreme spring drought (ESD) frequency differs significantly. The results of Poisson regression on ESD and PCs showed good predictive performance with R-squared value larger than 0.8. Furthermore, the results of applying the neural networks for PCs showed a significant improvement in the issue of under-estimation of the upper quartile group in ESD, with a high coefficient of determination of 0.91. This study identified PCs of large-scale ATPs that are candidate parameters for ESD prediction in the CDR. We expect that our findings can be helpful in undertaking mitigation measures for ESD in China.

Keywords: extreme spring drought; atmospheric teleconnection patterns; drought prediction; China

---

1. Introduction

Under global warming, the occurrence of extreme climate disasters is predicted to become more frequent, with droughts being one of the most severe disasters [1–4]. Annually droughts cause 6–8 billion dollars in damages globally, and people who are affected by droughts suffer far more damage than other natural disasters. The number of droughts has been increasing in China, which has experienced even more severe droughts over the past two decades [5–7].

Although China occupies 6% of the freshwater in the world, a total volume of 2.8 trillion m^3, the water availability per capita is only 25% of the global average. Furthermore, uneven spatial-temporal distribution of precipitation has been observed in recurrent droughts in northern China, as well as frequent floods in southern China in the summer, and reverse signals during the spring [1]. Significant economic and societal losses have been incurred due to persistent severe droughts in 1997 and from 1999 to 2002 in northern China [2]. In 2000, droughts affected more than 40 million hectares' agricultural areas [5]. The spring drought in 2011 over southern China has received much attention [6,8,9], as it resulted in more than 1.09 billion dollars in economic losses solely in Hubei Province. For the entire affected area, the destroyed crops exceeded 98.9 million hectares, and more than 4.9 million people and 3.4 million livestock suffered from deficient drinking water. This disaster has strongly impacted agriculture, industry, fisheries, and the daily life of the public. In recent decades, drought has attracted the attention of both the government and the public, and thus, the drying trend in China has been widely examined: Drought areas revealed an expanding trend in northern China in the last half-century using the China-Z index [7], while a drying signal in North China has also been detected using the surface humidity index during 1951–1998.

Drought management has evolved from passive management methods to proactive management methods, including overall risk analysis and hazard assessment [10]. However, it is challenging to define drought precisely due to its different temporal-spatial distribution and water demand of human society. Nowadays most drought studies are based on Palmer Drought Severity Index (PDSI) [11] using soil water balance equation or Standardized Precipitation Index (SPI) [12] using precipitation. Although PDSI can measure both wetness and dryness with statistical accuracy [11,13], SPI is preferred in this study due to its flexible temporal scale. Drought forecasting has been widely studied using both statistical and dynamical approaches [14]. Statistical methods are based on historical records, such as artificial neural networks and extreme learning machines [15]. Dynamical approaches refer to climatological or hydrological models based on variables (e.g., air temperature and precipitation) from atmospheric-ocean general circulation models (GCMs). Although well-designed statistical methods perform better in China, physical processes that are highlighted by dynamic models are often ignored [16].

As multiple causes generate droughts [17], it is recommended to examine the impact of atmospheric circulation anomalies [18] because of their protracted feedback on climate [19], which can be used to predict drought. As the spring drought in 2011 over southern China caused substantial social and economic losses, the causes of this drought have been widely investigated to avoid such great economic loss by accurately forecasting drought. Atmospheric circulation patterns have been found to exert significant influence in this event—the La Niña event in 2010–11 decreased the moisture transfer from tropical oceans to southern China and strengthened the dry flow from Siberia to China, while the positive North Atlantic Oscillation (NAO) in January–May 2011 also led to drier conditions in China through sub-polar and subtropical waveguides [4]. Jin et al. [9] reveal that the persistent spring drought events in 2011 were caused by maintenance of the quasi-stationary waves at middle-high latitudes and anomalous thermal-forcing over southeast Asia, which is possibly related to La Niña events. Zhang et al. [20] suggested that the western Pacific subtropical high may have a significant influence on drought patterns in China. Wu et al. [21] analyzed the relationship between recent drought events in Southwest China and anomalous atmospheric circulation patterns and stressed the need to detect the influencing factors, such as the Arctic Oscillation (AO). Although there are also apparent changes in spring rainfall, the mechanism for spring drought has received little attention and remains to be evaluated. Xin et al. [22] found that NAO is inversely correlated with the late spring upper-troposphere temperature with a three-month lag time over central China, and the inter-decadal change of winter NAO may cause the decrease in South China late spring rainfall. Kim et al. [23] found a three-month lagged correlation between NAO and spring precipitation in Korea and suggested that NAO be a possible parameter for drought forecasting. Understanding the atmospheric circulation patterns (ACPs) influencing drought development is essential in the management of this spatially extensive and recurrently prolonged natural hazard [24].

Therefore, the purpose of this study was to investigate the spatial-temporal characteristics of extreme droughts and to improve the prediction of extreme droughts related to large-scale atmospheric teleconnection patterns over China. The following questions were considered in this study: 1) What are the spatial-temporal characteristics of extreme droughts in China? 2) What is the association of extreme droughts linked with large-scale atmospheric teleconnection patterns across China? 3) How can we predict extreme droughts in the core drought region in China?

2. Data and Method

The daily precipitation data used in this study are complete data from 1961 to 2017 from 537 stations over China (Figure 1a shows the location of stations and Chinese topography), provided by the China Meteorological Data Service Center. Figure 1b shows the annual precipitation (mm) and nine major river basins in China. In this study, a seasonal Standardized Precipitation Index (SPI) is estimated to detect extreme drought conditions (SPI < −1.5).

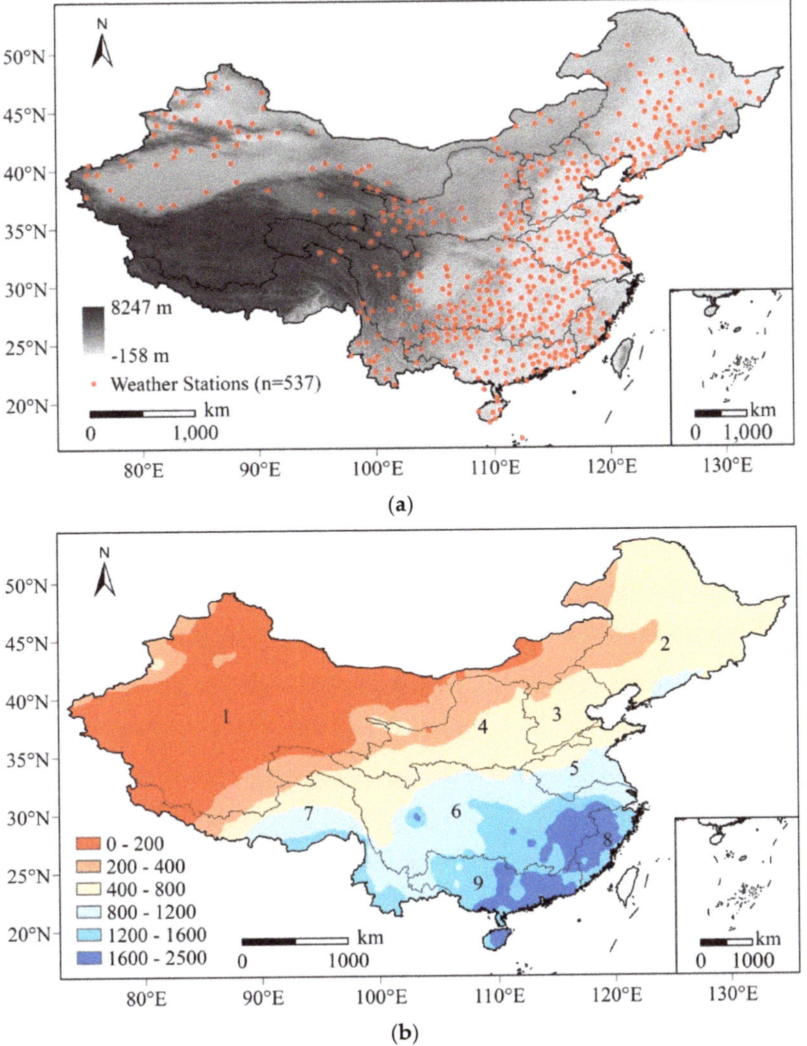

Figure 1. Precipitation data from 537 stations over China during the period 1961–17. (**a**) Location of 537 weather stations and topography over China; and (**b**) Annual precipitation (mm) and nine major river basins in China (1 = Continental River Basin; 2 = Songliao River Basin; 3 = Haihe River Basin; 4 = Yellow River Basin; 5 = Huaihe River Basin; 6 = Yangtze River Basin; 7 = Southwest River Basin; 8 = Southeast River Basin; 9 = Pearl River Basin).

The National Oceanic and Atmospheric Administration (NOAA; http://www.noaa.gov/) uses rotated principal component analysis to analyze monthly mean standardized 500 mb height anomalies in the region 20 °N to 90 °N and provides Northern Hemisphere teleconnection patterns as ten indices. Simultaneous and lagged correlations with atmospheric teleconnection indices were analyzed to select the candidate predictors of extreme drought in China: North Atlantic Oscillation (NAO), East Atlantic

(EA), East Atlantic/Western Russia (EA-WR), Scandinavia (SCA), Polar/Eurasia (POL), West Pacific (WP), Pacific/North American (PNA) and Tropical/Northern Hemisphere (TNH), as shown in Table 1.

Table 1. Major Atmospheric Teleconnection Patterns in Different Regions of the Northern Hemisphere.

Regions	Prominent Patterns
North Atlantic	North Atlantic Oscillation (NAO)
	East Atlantic (EA)
Eurasia	East Atlantic/Western Russia (EATL/WRUS)
	Scandinavia (SCA)
	Polar/Eurasia (POL)
North Pacific/North America	West Pacific (WP)
	Pacific/North American (PNA)
	Tropical/Northern Hemisphere (TNH)

One of the most important teleconnection patterns [25] is NAO, which is a north-south dipole (Greenland and central latitudes of the North Atlantic between 35°N and 40°N) of anomalies and shows considerable inter-annual or multi-decadal variability [26]. Several studies suggest that the preceding winter NAO will affect the early spring temperature in East Asia [27]. Sun and Yang [15] also find that the positive NAO might arouse an anomalous anti-cyclonic circulation over the North Atlantic and western Europe regions and lead to severe drought in southern China in spring 2011.

EA is the second prominent pattern of low-frequency variability across North Atlantic which is structurally similar to the NAO and consists of a north-south dipole of anomaly centers displaced southeastward to the approximate nodal lines of the NAO mode. EA/WR is one of three prominent teleconnection patterns affecting Eurasia all over the year and consists of four main anomaly centers. The main surface temperature anomalies linked with positive EA/WR reflect above-normal temperatures in East Asia and below-normal temperatures across western Russia and northeastern Africa. SCA consists primarily of circulation center in Scandinavia, with weaker centers across western Europe and eastern Russian/west Mongolia. POL exists whole the year and is linked with fluctuations in the intensity of the circumpolar circulation. WP is a primary pattern of low-frequency variability in North Pacific consisting of a north-south dipole of anomalies [25] (the Kamchatka Peninsula and portions of southeastern Asia and the western subtropical North Pacific) during winter and spring. PNA is an important pattern of low-frequency variability in the Northern Hemisphere extra-tropics. Although PNA is a natural internal pattern of climate variability, it is also strongly influenced by the El Niño/Southern Oscillation (ENSO) phenomenon. TNH was first proposed by Mo and Livezey [28] and exists as a wintertime pattern.

Trend analysis is often used to detect changes in time-series data. One of the commonly used non-parametric trend tests is called the Mann–Kendall (MK) test [21,22,29,30]. The null hypothesis H_0 indicates that sample data (X_i, $i = 1, 2 \ldots , n$) are randomly ordered and there is no trend, while the alternative hypothesis H_1 represents a monotonic trend in X. MK test uses Z to denote the trend in time-series data: The symbol of Z represents the trend direction (positive for an upward trend, vise-versa for negative), and the magnitude of Z relates to the slope of trend and the significance level, where $|Z| > 1.64$ for the 10% significance level and $|Z| > 1.96$ for the 5% significance level. Because the existence of autocorrelation in the dataset will affect the probability of detecting trends, which is often ignored, Hamed and Rao [31] developed a modified non-parametric trend test whose accuracy is better than the original MK test when considering empirical significance level. In this study, the modified MK test is adopted.

For multiple possible prediction parameters, we first extracted several principal components (PCs) using principal component analysis (PCA) [32], a function-reduction method, to account for a large part of the variance of original parameters and to predict decadal extreme spring droughts (ESDs). The tree-based model [8,33] is then used to classify extreme droughts (response variables) in spring based on the PCs (potential precursors). Poisson regression, which is often used for prediction of count data, as well as neural networks are adopted for predicting ESDs. Neural networks, which is more often called artificial neural networks (ANN), are nonlinear models that can discover data patterns and estimate complex relationship with high accuracy [2]. Among many types of neural networks, a feed-forward ANN comprising of multiple neurons arranged in layers is the most popular model for time-series forecasting [32]. ANN is typically made up of several layers of nodes. Input and output layers receive external information and export results. One or more intermediate layers contains the complex calculation processes, which is often described as a black box. Next, Receiver Operating Characteristics (ROC) analysis, which was initially developed during World War II for classifying radar signals from noise [34], is used to test the results of Poisson regression and neural networks. ROC analysis is typically used to evaluate the performance of a statistical classification model based on the sensitivity (TPR, true positive rate) and 1-specificity (FPR, false positive rate), which can be estimated by:

$$TPR = \frac{TP}{TP + FN}; FPR = \frac{FP}{FP + TN} \tag{1}$$

where TP indicates that the predicted and original results are both true, FP indicates that the predicted result is true, but the original result is false, TN indicates that the predicted and original results are both false, and FN indicates that the predicted result is false, but the original result is true. More details about ROC are described by Zou et al. [34].

3. Results

3.1. Diagnostic Analysis of Historical Spring Droughts in China

In this study, three representative droughts (2001/2007/2011) were selected to assess the drought development process in accordance with SPI3 and the composite anomalies of the meridional spring wind (m/s) conditions between 850 mb and 250 mb in the corresponding area over China (Figure 2). The upper three figures give drought information using SPI3, while the lower three figures show meridional wind condition. The range of latitude in the lower figures corresponds to extreme drought region in the upper figure, and the range of longitude accords with the upper map, the y-axis of lower figures ranges from 850 mb to 250 mb. Red color in the upper figures denotes drought condition and blue color in the lower figures represents weaker meridional wind and vice versa.

In 2001, nearly all stations over north and central China undergone a drought condition (SPI < −0.5), while many stations, covering the entire Huaihe river basin, northwest of the Yangtze river basin, southeast of the Yellow river basin and south of Haihe/Songliao river basin, have experienced an extreme drought (SPI < −1.5), as shown in Figure 2a1. Figure 2a2 displays the meridional wind composite anomaly between 25 °N–40 °N and 70 °E–135 °E. There are visible blue band ranging from 70 °E–90 °E in the higher atmosphere and 100 °E–125 °E in the lower atmosphere. As there are limited meteorological stations and the effect of meridional wind is also weaker on local weather due to high altitude over China in the region 70 °E–90 °E, the significant influence of weaker meridional wind that can also be detected comes from the range 100 °E–125 °E, where extreme drought prevails. The mechanism that weaker meridional wind led to lower precipitation is also apparent—less moisture is transferred by meridional wind from the south sea to land in the north.

In 2007, a drought disaster struck southern China, including south of Yangtze river basin, Pearl river basin, and Southeast river basin, where the population is densely distributed, and the economy is very prosperous, and drought can cause much more significant social and economic loss. Although approximately only 20 stations are undergoing extreme drought, the large spatial scale of this disaster also appeals for public attention. In the corresponding longitude (100 °E–120 °E), blue color prevails in the meridional wind graph. Although there was stronger meridional wind in the upper atmosphere, its influence on drought appears to be weaker due to thin air.

As mentioned before, an extreme drought occurred during the spring of 2011, which is displayed by Figure 2c1. Large parts of Yangtze river basin and part of Huaihe/Southeast/Pearl river basin suffer striking extreme drought. Figure 2c2 shows the composite anomaly of meridional wind at 20 °N–40 °N and 70 °E–130 °E. Over the drought region, the meridional winds were weaker than normal throughout the entire atmospheric layer. This strong correlation may be a useful signal to predict extreme drought over China.

From the investigation of historical droughts, a core drought region (CDR) with range (25 °N–35 °N, 105 °E–120 °E) under more severe drought risk is selected for the forecasting of ESD. There are, in total, 176 stations within CDR, 120 stations (68%) in the lower reach of the Yangtze river basin, 27 stations (15%) in the Huaihe river basin, and 13/11/4 stations in Southeastern/Pearl/Yellow river basins, respectively.

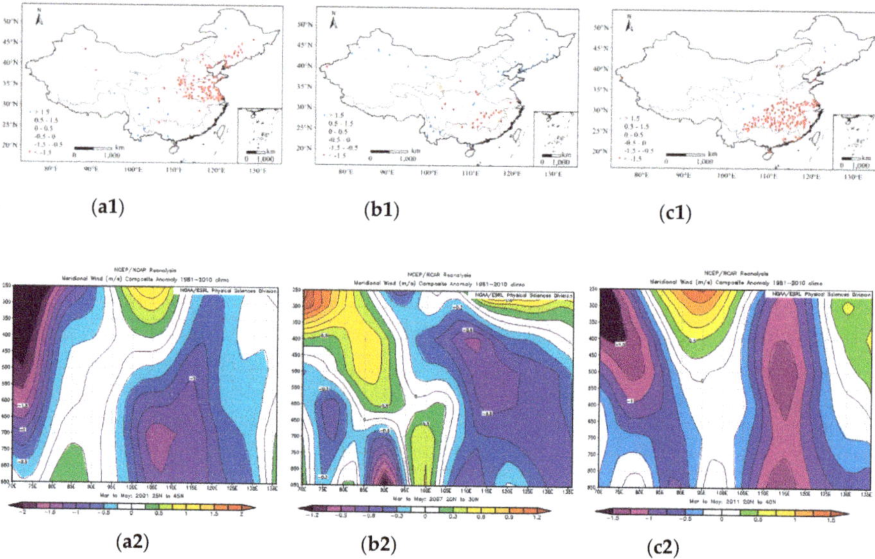

Figure 2. Spring extreme droughts and composite anomaly of meridional wind (m/s) at the corresponding areas in 2001 (**a1/a2**), 2007 (**b1/b2**), and 2011(**c1/c2**).

3.2. Changes in Spring Precipitation and Extreme Droughts in China

The trend of spring precipitation and extreme drought is shown in Figure 3. The shift in spring precipitation indicates a significant decadal drying trend in central China (125 stations with 10% significance and 103 stations with 5% significance), which matches the CDR well. In CDR, there are 68 stations (53%) and 56 stations (43%) showing a decrease in spring precipitation with 10% and 5% significance, respectively. Although northeastern China has received more rainfall, the major water resources in China come from precipitation in central and southeast China, where the population is densely distributed, and industry is relatively more prosperous. As central China is more water-demanding and undergoes a decrease in spring rainfall, it is vital to study the influence of decadal water resource reduction on water resource management in this region. Compared to the trend of spring precipitation, a smaller area around the lower reach of the Yangtze river basin suffers more frequent ESD, where 43 (33%) and 24 (19%) stations in CDR show an increase in ESD frequency at 10% and 5% significance levels, respectively. Owing to the enormous economic losses caused by drought disasters, the cause and prediction methods of ESD are studied in this paper to promote better mitigation measures.

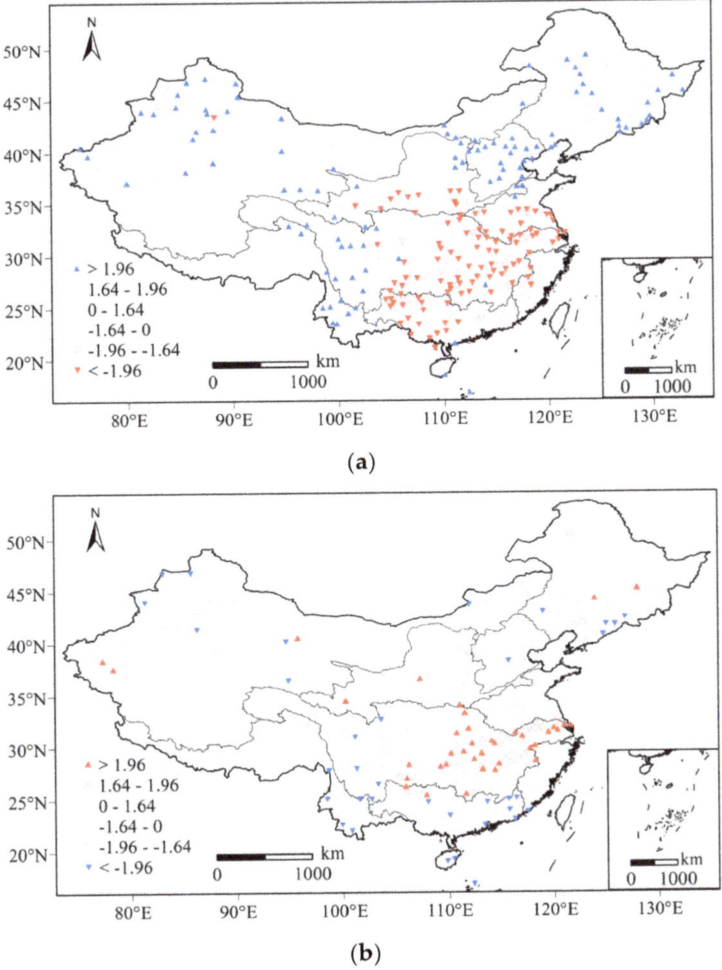

Figure 3. (**a**) Modified Mann–Kendall trend of spring precipitation; and (**b**) extreme drought frequency at 10% and 5% significance levels.

3.3. Extreme Spring Drought Prediction in the CDR

PCA has been applied on the eight possible precursors of ESD to obtain linearly uncorrelated variables (PCs). The dimension reduction result is shown in Table 2. Although there are many rules to select PCs—based upon eigenvalues of the covariance matrix of dataset or proportion of variance—we select PCs according to the requirement and result of the following analysis. Based on PC1 and PC2, we use a tree-based model to classify ESD frequency in CDR into three groups, as shown in Figure 4a: Group 1, PC1 $\geq$ 0.04479; Group 2, PC1 < 0.04479 and PC2 < $-$0.02289; Group 3, PC1 < 0.04479 and PC2 $\geq$ $-$0.02289. From the empirical probability density of ESD frequency of each group in Figure 4b, Group 3 shows a higher probability in higher ESD frequency. Figure 4c uses PC1 and PC2 to partition ESD, where blue background means Group 1 and red for Group 3. The red triangles and blue inverted triangles represent upper and lower quantile of ESD, with all red triangles located in Group 3 and all blue inverted triangles in Group 1. That is, when PC1 < 0. 04479 and PC2 $\geq$ −0.02289, the ESD frequency is significantly greater. This result may be used as an effective warning signal—when PC1 < 0.04479 and PC2 $\geq$ −0.02289, the drought risk is higher over CDR.

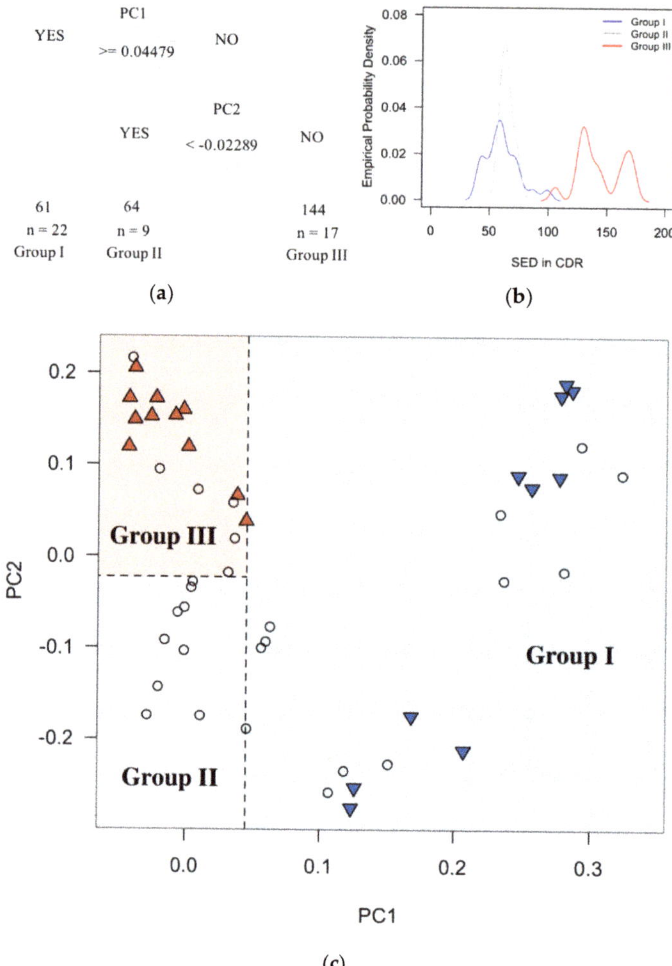

Figure 4. Classification of decadal spring extreme drought frequency in the core drought region based on (**a**) tree-based model; (**b**) probability density of each group; and (**c**) principal component 1 (PC1) and principal component 2 (PC2).

Table 2. Explained variance of principal components (PCs) for decadal teleconnection indices.

Importance of Components	PC1	PC2	PC3	PC4	PC5	PC6	PC7	PC8
Standard deviation	29.6473	12.3784	9.96176	5.87891	4.31489	3.18466	2.01416	3.378×10^{-15}
Proportion of Variance	0.7332	0.1278	0.08278	0.02883	0.01553	0.00846	0.00338	0.00001
Cumulative Proportion	0.7332	0.861	0.94379	0.97262	0.98816	0.99662	0.99999	1

ESD frequency is count data and shows a strong correlation with PCs. Consequently, we used Poisson regression and neural networks to predict ESD. After trying different options to select PCs, we choose PC1~4 as prediction parameters, which gives much better results than PC1~3 and cannot be significantly improved by additionally choosing PC5. Table 3 shows the coefficients of each possible predictors and the corresponding p-values (all p-values are near to 0) using Poisson regression, while the coefficient of determination (R-squared value) between original ESD and simulated results by Poisson regression is 0.834, which strongly suggests these variables to be prediction parameters.

Table 3. Coefficients of Poisson regression.

	Estimate	Std. Error	z value	$Pr(>|z|)$	Remark
Intercept	4.6318	0.0186	249.018	$<2 \times 10^{-16}$	***
PC1	−2.3031	0.1483	−15.530	$<2 \times 10^{-16}$	***
PC2	1.5252	0.1111	13.728	$<2 \times 10^{-16}$	***
PC3	−1.0643	0.1128	−9.437	$<2 \times 10^{-16}$	***
PC4	−0.7187	0.1155	−6.220	5×10^{-10}	***

*** means significance level is nearly 0.

After testing the effectiveness of four candidate predictors, we used leave-one-out cross-validation to investigate the predictability of Poisson regression and neural networks. The predicted ESD versus original records is given in Figure 5a,b, where the high R-squared value (0.791 for Poisson regression and 0.924 for neural networks) suggests PCs of teleconnection indices to be useful in the prediction of ESD over CDR. In addition, the application of the neural networks to the PCs resulted in a significant improvement in the under-estimation of the top quartile group in ESD with a high determinant ($R^2 = 0.924$). For the ROC analysis, we predict the ESD of one decade for each time using Poisson regression and neural networks, and then classify the predicted result and original ESD based on thresholds, ranging from minimum to maximum ESD, to be positive (both greater than, equal to or less than the threshold) or negative. After that, we plot the results as shown in Figure 5c,d. The diagonal line connecting the left bottom and top right corners represent random guess results. Points below the line indicate worse results than random guesses, points closer to (0,1) suggest better results. The red points in the figure indicate FPR < 0.1 and TPR > 0.8, so the corresponding thresholds (around 80 for Poisson regression and 100 for neural networks) are useful in classifying the predicted results. For neural networks, ten thresholds around 100 can lead to a perfect classification result with FPR = 0 and TPR = 1. This result suggests Poisson regression and neural networks to be helpful in predicting ESD over CDR using teleconnection indices.

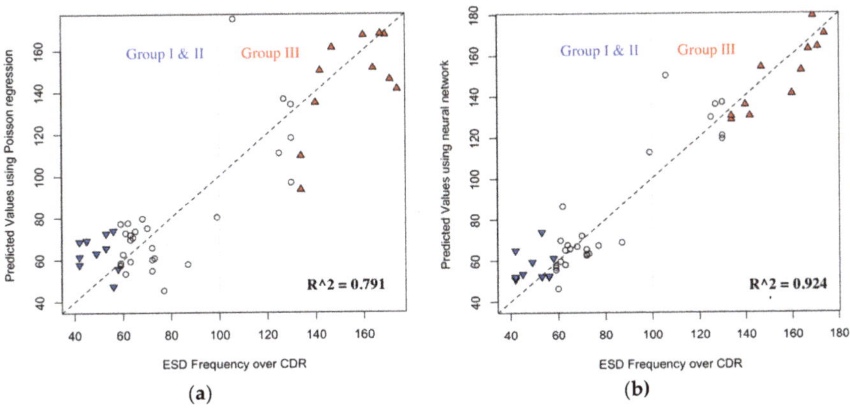

Figure 5. *Cont.*

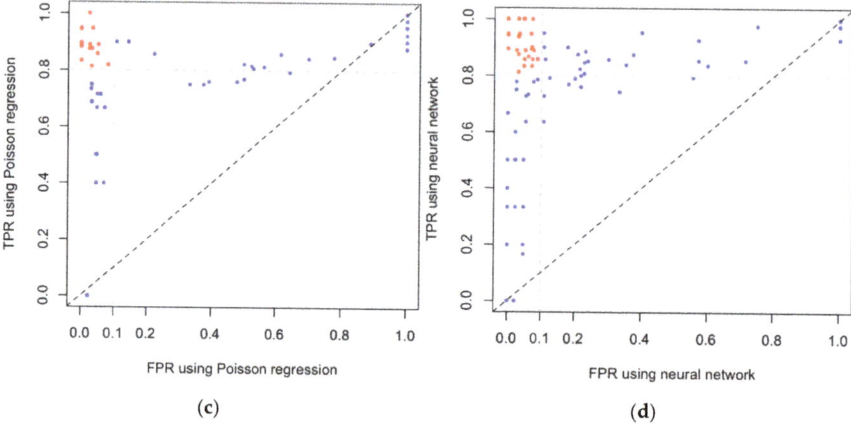

Figure 5. Extreme spring drought (ESD) frequency over CDR (**a**) using Poisson regression; and (**b**) neural networks. Receiver Operating Characteristics (ROC) evaluation for prediction results of (**c**) Poisson regression; and (**d**) neural networks.

4. Summary and Conclusions

This study analyzed the trend of spring precipitation and ESD frequency and their lagged correlation with atmospheric teleconnection patterns over China to detect the characteristics and causes of ESD and improve its predictability. Based on complete daily precipitation data from 1961–2017 covering 537 stations over Chia, SPI3 are computed for drought analysis. The main results are summarized as follows. In recent decades (2001/2007/2011), central and southeast China has undergone several severe drought disasters. Therefore, a CDR over 25 °N–35 °N, 105 °E–120 °E under more severe drought risk is defined for drought prediction in this study. According to modified MK test on spring precipitation and ESD, it was found that there is a significant decadal drying trend in central and southeast China, which is approximately identical with CDR. To predict drought events over CDR, eight teleconnection indices have been chosen as candidate parameters through lagged correlation analysis. Firstly, PCA is employed to extract useful information in predictors and reserve it in linearly uncorrelated variables, and then these PCs are adapted to conduct further forecasting analysis.

Based on PC1 and PC2 using the tree-based model, ESD over CDR is classified into three groups, where ESD frequency is higher in Group 3 according to the empirical probability function. This result may be used as a useful warning signal—when PC1 < 0. 04479 and PC2 $\geq$ −0.02289, the drought risk is higher over CDR. When we use Poisson regression to model ESD over CDR, the coefficient of determination between the model result and original ESD has reached 0.834, which strongly suggests these variables to be prediction parameters.

ESD over CDR is well predicted by Poisson regression and neural networks using PCs, with the R-squared value equal to 0.834 and 0.914. The ROC evaluation of the prediction results also suggests these two models be candidate prediction techniques. In addition, this study identified the applicability of the ESD prediction in China's CDR through combined neural networks and PCs for ATPs. The results of the study were based on limited observations, but this study makes it possible to diagnose ESD and clarify its drought prediction techniques and practical strategies for coping with ESD in China. Then Leave-one-out cross-validation is applied to investigate the predictability of Poisson regression and neural networks. The higher R-squared value 0.924 for neural networks (compared to 0.791 for Poisson regression) reveals that ESD over CDR is highly predictable using PCs and appropriate methodology. Besides, the application of the neural networks to the PCs resulted in a significant improvement in the under-estimation of the top quartile group in ESD. Using the ROC analysis, ten thresholds around 100 can lead to a perfect classification result with FPR = 0 and TPR = 1 based on neural networks. Based on the above analysis, we highly suggest that the excellent classification result of ESD based on PC1 and PC2 and the high predictability using neural networks on ESD may be helpful in drought management in central and southeast China.

Author Contributions: conceptualization, Q.G. and J.-S.K.; Formal analysis, Q.G. and J.-S.K.; Methodology, J.C. and J.-S.K.; resources, H.C.; writing—original draft preparation, Q.G. and J.-S.K.; writing—review and editing, J.C., H.C., and J.-H.L.

Funding: This collaborative research was funded by the National Research Foundation of Korea (grant number: NRF-2018K2A9A06024031) and the National Natural Science Foundation of China (grant number: NSFC-51811540407).

Conflicts of Interest: The authors declare no conflicts of interest.

References

1. Jefferson, M. *IPCC Fifth Assessment Synthesis Report: "Climate Change 2014: Longer Report": Critical Analysis*; Elsevier: New York, NY, USA, 2015; pp. 362–363.

2. Mishra, A.K.; Desai, V.R. Drought forecasting using feed-forward recursive neural network. *Ecol. Model.* **2006**, *198*, 127–138. [CrossRef]

3. Chen, S.; Muhammad, W.; Lee, J.H.; Kim, T.W. Assessment of Probabilistic Multi-Index Drought Using a Dynamic Naive Bayesian Classifier. *Water Resour. Manag.* **2018**, *32*, 4359–4374. [CrossRef]

4. Lee, J.H.; Park, S.Y.; Kim, J.S.; Sur, C.Y.; Chen, J. Extreme drought hotspot analysis for adaptation to a changing climate: Assessment of applicability to the five major river basins of the Korean Peninsula. *Int. J. Climatol.* **2018**, *38*, 4025–4032. [CrossRef]

5. Dai, A. Increasing drought under global warming in observations and models. *Nat. Clim. Chang.* **2013**, *3*, 52–58. [CrossRef]

6. Dai, A. Drought under global warming: A review. *Wiley Interdiscip. Rev. Clim. Chang.* **2011**, *2*, 45–65. [CrossRef]

7. Zhai, J.Q.; Su, B.; Krysanova, V.; Vetter, T.; Gao, C.; Jiang, T. Spatial variation and trends in PDSI and SPI indices and their relation to streamflow in 10 large regions of China. *J. Clim.* **2010**, *23*, 649–663. [CrossRef]

8. Breiman, L. *Classification and Regression Trees*; Routledge: Abingdon, UK, 2017.

9. Jin, D.; Guan, Z.; Tang, W. The Extreme Drought Event during Winter–Spring of 2011 in East China: Combined Influences of Teleconnection in Midhigh Latitudes and Thermal Forcing in Maritime Continent Region. *J. Clim.* **2013**, *26*, 8210–8222. [CrossRef]

10. Zhang, Q. Drought and its impacts. In *China Climate Impact Assessment*; China Meteorol Press: Beijing, China, 2003.

11. Palmer, W.C. *Meteorological Drought, Weather Bureau Research Paper No. 45*; United States Department of Commerce: Washington, DC, USA, 1965.

12. McKee, T.B.; Doesken, N.J.; Kleist, J. The relationship of drought frequency and duration to time scales. In Proceedings of the 8th Conference on Applied Climatology, Boston, MA, USA, 17–22 January 1993; American Meteorological Society: Boston, MA, USA, 1993; pp. 179–183.

13. Heim, J.R.; Richard, R. A review of twentieth-century drought indices used in the United States. *Bull. Am. Meteorol. Soc.* **2002**, *83*, 1149–1165. [CrossRef]

14. Zou, X.; Zhai, P.; Zhang, Q. Variations in droughts over China: 1951–2003. *Geophys. Res. Lett.* **2005**, *32*, L04707. [CrossRef]

15. Sun, C.; Yang, S. Persistent severe drought in southern China during winter–spring 2011: Large-scale circulation patterns and possible impacting factors. *J. Geophys. Res. Atmos.* **2012**, *117*. [CrossRef]

16. Wang, Z.; Zhai, P. Climate Change in Drought over Northern China during 1950–2000. *Acta Geogr. Sin.* **2003**, *58*, 61–68.

17. Hayes, M.J.; Wilhelmi, O.V.; Knutson, C.L. Reducing drought risk: Bridging theory and practice. *Nat. Hazards Rev.* **2004**, *5*, 106–113. [CrossRef]

18. Mishra, A.K.; Singh, V.P. A review of drought concepts. *J. Hydrol.* **2010**, *391*, 202–216. [CrossRef]

19. Deo, R.C.; Şahin, M. Application of the extreme learning machine algorithm for the prediction of monthly Effective Drought Index in eastern Australia. *Atmos. Res.* **2015**, *153*, 512–525. [CrossRef]

20. Zhang, Z.; Jin, Q.; Chen, X.; Xu, C.Y.; Jiang, S. On the linkage between the extreme drought and pluvial patterns in china and the large-scale atmospheric circulation. *Adv. Meteorol.* **2016**, *2016*, 8010638. [CrossRef]

21. Wu, Y.; Wu, S.; Wen, J.; Xu, M.; Tan, J. Changing characteristics of precipitation in China during 1960–2012. *Int. J. Climatol.* **2016**, *36*, 1387–1402. [CrossRef]

22. Xin, X.; Yu, R.; Zhou, T.; Wang, B. Drought in Late Spring of South China in Recent Decades. *J. Clim.* **2006**, *19*, 3197–3206. [CrossRef]

23. Kim, J.S.; Seo, G.S.; Jang, H.W.; Lee, J.H. Correlation analysis between Korean spring drought and large-scale teleconnection patterns for drought forecasting. *KSCE J. Civ. Eng.* **2017**, *21*, 458–466. [CrossRef]

24. Irannezhad, M.; Chen, D.; Kløve, B.; Moradkhani, H. Analysing the variability and trends of precipitation extremes in Finland and their connection to atmospheric circulation patterns. *Int. J. Climatol.* **2017**, *37*, 1053–1066. [CrossRef]

25. Barnston, A.G.; Livezey, R.E. Classification, seasonality and persistence of low-frequency atmospheric circulation patterns. *Mon. Weather Rev.* **1987**, *115*, 1083–1126. [CrossRef]

26. Hurrell, J.W. Decadal trends in the North Atlantic Oscillation: Regional temperatures and precipitation. *Science* **1995**, *269*, 676–679. [CrossRef] [PubMed]

27. Yu, R.; Zhou, T. Impacts of winter-NAO on March cooling trends over subtropical Eurasia continent in the recent half century. *Geophys. Res. Lett.* **2004**, *31*, 261–268. [CrossRef]

28. Mo, R.E.; Livezey, R.E. Tropical-extratropical geopotential height teleconnections during the Northern Hemisphere winter. *Mon. Weather Rev.* **1986**, *114*, 2488–2515. [CrossRef]

29. Myronidis, D.; Stathis, D.; Ioannou, K.; Fotakis, D. An Integration of Statistics Temporal Methods to Track the Effect of Drought in a Shallow Mediterranean Lake. *Water Resour. Manag.* **2012**, *26*, 4587–4605. [CrossRef]

30. Szinell, C.S.; Bussay, A.; Szentimrey, T. Drought tendencies in Hungary. *Int. J. Climatol. J. R. Meteorol. Soc.* **1998**, *18*, 1479–1491. [CrossRef]

31. Hamed, K.H.; Rao, A.R. A modified Mann-Kendall trend test for autocorrelated data. *J. Hydrol.* **1998**, *204*, 182–196. [CrossRef]

32. Wold, S.; Esbensen, K.; Geladi, P. Principal component analysis. *Chemom. Intell. Lab. Syst.* **1987**, *2*, 37–52. [CrossRef]

33. Kim, J.S.; Jain, S.; Moon, Y.I. Atmospheric teleconnection-based conditional streamflow distributions for the Han River and its sub-watersheds in Korea. *Int. J. Climatol.* **2012**, *32*, 1466–1474. [CrossRef]

34. Zou, K.H.; O'Malley, A.J.; Mauri, L. Receiver-operating characteristic analysis for evaluating diagnostic tests and predictive models. *Circulation* **2007**, *115*, 654–657. [CrossRef]

Article

Stochastic Model for Drought Forecasting in the Southern Taiwan Basin

Hsin-Fu Yeh * and Hsin-Li Hsu

Department of Resources of Engineering, National Cheng Kung University, Tainan 701, Taiwan; pig5206332@gmail.com
* Correspondence: hfyeh22@gmail.com; Tel.: +886-6-275-7575

Received: 1 August 2019; Accepted: 28 September 2019; Published: 29 September 2019

Abstract: The global rainfall pattern has changed because of climate change, leading to numerous natural hazards, such as drought. Because drought events have led to many disasters globally, it is necessary to create an early warning system. Drought forecasting is an important step toward developing such a system. In this study, we utilized the stochastic, autoregressive integrated moving average (ARIMA) model to predict drought conditions based on the standardized precipitation index (SPI) in southern Taiwan. We employed data from 1967 to 2006 to train the model and data from 2007 to 2017 for model validation. The results showed that the coefficients of determination (R^2) were over 0.80 at each station, and the root-mean-square error and mean absolute error were sufficiently low, indicating that the ARIMA model is effective and adequate for our stations. Finally, we employed the ARIMA model to forecast future drought conditions from 2019 to 2022. The results yielded relatively low SPI values in southern Taiwan in future summers. In summary, we successfully constructed an ARIMA model to forecast drought. The information in this study can act as a reference for water resource management.

Keywords: stochastic model; ARIMA model; drought forecasting; southern Taiwan

---

1. Introduction

Drought is a natural phenomenon which can be classified into four categories: meteorological, hydrological, agricultural, and socioeconomic drought. Among these, meteorological drought can precede and lead to the other three types. It is essentially caused by insufficient precipitation and usually results in severe disaster events. Dai [1] found that the global trends of observed annual precipitation from 1950 to 2010 showed a decline in areas such as Africa, Southeast and East Asia, Eastern Australia, and Southern Europe. These results indicate that water shortages are becoming increasingly severe, which may lead to drought. However, drought events not only lead to environmental changes but also have serious impacts on the social economy and agricultural development [2]. For example, between 1976 and 2006, the total cost associated with drought in Europe was as high as €100 billion, while the number of people affected by drought increased by 20% [3]. Therefore, it is important to assess drought conditions to develop a mitigation strategy.

Many drought indices have been proposed to assess meteorological drought to identify the occurrence of drought events. Among these, the standardized precipitation index (SPI) has been widely employed to evaluate the intensity of meteorological droughts, owing to the following advantages. First, the SPI is based only on long-term rainfall data. Second, the SPI is in a standardized form, which means that it can be compared between different regions. Third, it has variable timescales, which allows it to describe different drought conditions. Therefore, the SPI was selected as the meteorological drought index in the present study.

Because a drought varies slowly in time, it is generally not discovered until a disaster occurs. Therefore, an effective monitoring system is required to mitigate the impacts of droughts. An effective monitoring system can help to establish an early warning system for drought [4], and drought forecasting is an important step toward developing such an early warning system. Forecasting future drought conditions in a region is critical for the sustainability of water resource planning and management. In addition, drought forecasting plays a vital role in drought risk assessment [5,6] and provides information for water resource managers and policy makers to take precautions against droughts in advance [7]. There have been many previous studies concerning drought forecasting, such as detecting occurrences of drought and predicting the duration and severity of a drought event. However, the main challenge is to select a suitable and effective tool for forecasting [6–8].

There exist many methodologies for forecasting drought events based on drought indices, such as regression analysis [9,10], stochastic models [7,11–14], probability models [15], artificial intelligence (AI)-based models [16–19], and dynamic modeling [20,21].

The stochastic model, also known as the time series model, has been utilized to predict hydrological time series. In general, the time series model has a weak capability for modeling data with nonlinear characteristics but is able to effectively fit linear data such as streamflow and precipitation. In addition, there are periodic characteristics and serial correlations between observations in hydrological time series, and the time series model can describe these features well and perform systematic modeling. The most useful and common class of stochastic model is the autoregressive integrated moving average (ARIMA) model. The advantages of the ARIMA model include low data input requirements, a simple computational process [22], and few model parameters being required to describing the time series [10,11]. It is also suitable for nonstationary data. Owing to these merits, this time series model is superior to other statistical models [7,23].

The purpose of this study was to develop a valid linear stochastic model based on an SPI time series in southern Taiwan. According to a previous study, most studies have only developed the ARIMA model to fit their own time series data, and few studies have actually predicted future drought situations. Hence, in this study, we utilized the ARIMA model to forecast drought conditions over next four years, from 2019 to 2022.

2. Study Area

The study area is located in southern Taiwan and includes the Pozi River basin, Bazhang River basin, Jishui River basin, Yanshui River basin, Erren River basin, Gaopping River basin, and Linbian River basin. It belongs to the tropical monsoon climate zone. There is abundant rainfall, but owing to the uneven distribution of rainfall in time and space, the region is characterized by distinct wet and dry seasons. In this study, we selected one rainfall station in each basin, with records from the years 1967 to 2017. Information on the rainfall stations is presented in Table 1, and the spatial distributions of the rainfall stations at each basin are shown in Figure 1.

Table 1. Information on rainfall stations at each basin.

Basin	Station	Short Name
Pozi River basin	Zhang Nao Liao-2	PR1
Bazhang River basin	Da Hu Shan	BR1
Jishui River basin	Guan Zi Ling-2	GR1
Yanshui River basin	Qi Ding	YR2
Erren River basin	Gu Ting Keng	RR2
Gaoping River basin	Ping Dong-5	KR3
Linbian River basin	Nan Han	LR2

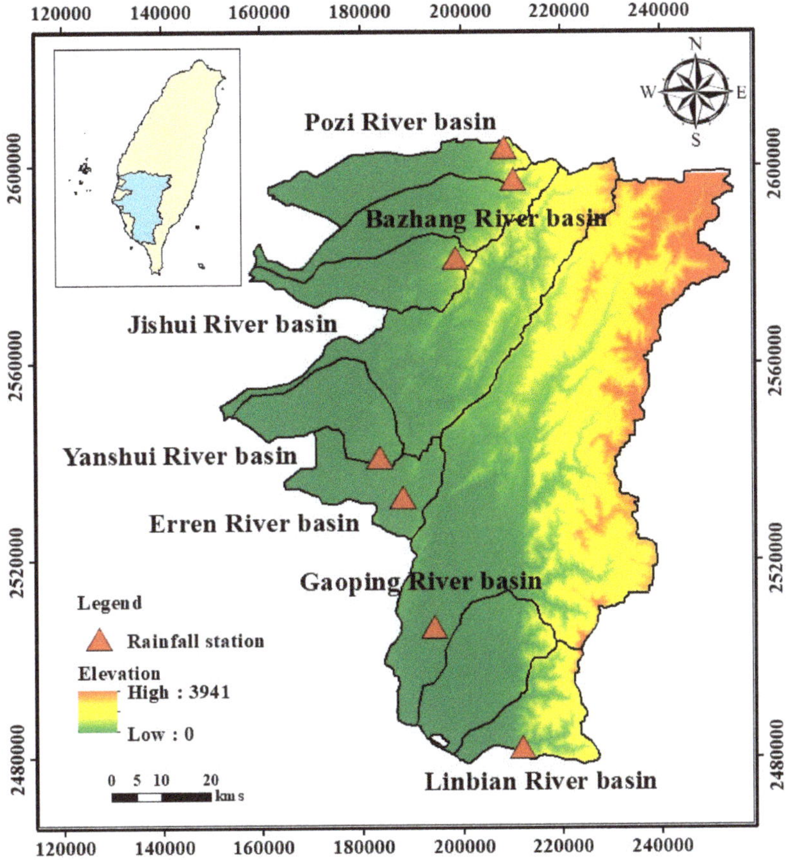

Figure 1. Spatial distribution of the elevation and rainfall stations at each basin.

3. Methodology

3.1. SPI

There exist many methods that can identify drought events, and using drought indicators is the most convenient and effective method to do this. In this study, we explored meteorological drought in southern Taiwan. The SPI was proposed by McKee et al. [24]. This index simply utilizes the cumulative precipitation data over different periods for calculation. Because the index is standardized, it can be compared between different regions. The World Meteorological Organization [25] listed this as the preferred index for describing precipitation drought events. Considering the rainfall pattern in Taiwan, over a shorter aggregation time scale, the SPI only exhibits periodic oscillations in time, and it is difficult to evaluate long-term drought events. In contrast, it is easier to identify drought events with a longer aggregation time scale. Therefore, we chose a 12 month aggregation time scale for the SPI. In this study, we utilized historical SPI time series to predict future drought conditions using a time series model.

3.2. Time Series Model

The time series model is a stochastic model that is commonly used for time series forecasting. Common time series models include the autoregressive (AR), moving average (MA), and autoregressive moving average (ARMA) models. Many studies have employed time series models to predict hydrological and meteorological time series [7,11,26].

3.2.1. Nonseasonal ARIMA Model

The AR and MA models can be effectively combined into an ARMA model. The current data is analyzed by a linear combination of previous data plus error terms. However, the ARMA model is only applicable when the data is stationary. If the original time series is nonstationary, then the difference must be added. The resulting model is called the ARIMA model, as proposed by Box and Jenkins [27]. The ARIMA model provides a new generation of forecasting tools, emphasizing the stochastic properties of time series rather than constructing single and simultaneous equation models. Each variable of the ARIMA model is represented by its own lagged value and stochastic error terms. The general nonseasonal ARIMA model consists of a p-order AR model, a q-order MA model, and the operators on the dth difference of the original time series data. This can be expressed as ARIMA(p,d,q), and the algorithm of the model is as follows:

$$\phi(L)\nabla^d x_t = \theta(L)\varepsilon_t \tag{1}$$

where $\phi(L)$ and $\theta(L)$ are polynomials of order p and q, respectively. The operators for the nonseasonal AR model of order p and MA model of order q are written as

$$\begin{aligned}\phi(L) &= (1 - \phi_1 L - \phi_2 L^2 - \cdots - \phi_p L^p) \\ \theta(L) &= (1 - \theta_1 L - \theta_2 L^2 - \cdots - \theta_p L^p)\end{aligned} \tag{2}$$

where L is the lag operator $L^i x_t = x_{t-i}$.

3.2.2. Seasonal ARIMA (SARIMA) Model

Box et al. [28] extended the ARIMA model to deal with seasonality, with the result commonly referred to as the SARIMA model. The SARIMA model is analyzed by introducing seasonal periodic change to a general ARIMA mode, and it can be denoted as ARIMA$(p,d,q)(P,D,Q)_S$, where (p,d,q) is the nonseasonal part of the model and $(P,D,Q)_S$ is the seasonal part. This can be expressed as follows:

$$\phi_p(L)\Phi_P(L^s)\nabla^d \nabla_s^D x_t = \theta_q(L^s)\Theta_Q(L^s)\varepsilon_t \tag{3}$$

where p is the order of the nonseasonal AR model, q is the order of the nonseasonal MA model, d is the order of the general difference, P is the order of the seasonal AR model, Q is the order of the seasonal MA model, D is the seasonal differencing, and S is the length of a season.

3.3. ARIMA Model Development

In general, the development and selection of an appropriate ARIMA model consists of three steps: model identification, parameter estimation, and diagnostic checking [23,27,29,30].

3.3.1. Model Identification

This step involves transforming the data which is to confirm the original data to normality and stationarity and utilizing the autocorrelation function (ACF) and partial autocorrelation function (PACF) to initially confirm the general form of the ARIMA model. Here, the ACF and PACF were used to determine the order of the model, and the final model was selected using the goodness-of-fit criteria through the Akaike information criterion (AIC) [31] and Schwarz–Bayesian criterion (SBC) [32]. The model that gives the minimum AIC and SBC value is selected as the best. The mathematical formulations of the AIC and SBC are as follows:

$$AIC = n \ln(MSE) + 2m \tag{4}$$

$$SBC = n \ln(MSE) + m \ln(n) \tag{5}$$

where $m = (p + q + P + Q)$ is the number of model parameters, n is the number of data, and MSE is the mean-square error.

$$\text{Mean square error } (MSE) = \frac{1}{n} \sum_{i=1}^{n} (y_t - \hat{y}_t)^2 \tag{6}$$

where y_t is the observed data and $\hat{y}_t$ is the predicted value.

3.3.2. Parameter Estimation

After identifying the ARIMA model, the parameters of the model must be estimated. This study employed the method of maximum likelihood, as proposed by Box and Jenkins [27], to estimate the parameters of the ARIMA model. Maximum likelihood estimation is a method of estimating the parameters of a statistical model.

3.3.3. Diagnostic Checking

Once an appropriate model has been selected and the parameters have been estimated, diagnosing the ARIMA model represents a crucial component of model development. In this step, it is necessary to verify whether the residuals of the model satisfy the properties of independence, following a normal distribution, and homoscedasticity to verify that the model is suitable for time series. Several statistical tests and plots of residuals are utilized for diagnostic checking. For a good model, the residuals must satisfy the white noise process requirements of being uncorrelated and normally distributed around a zero mean.

The residual autocorrelation function (RACF) and residual partial autocorrelation function (RPACF) of a time series are utilized to determine whether the series is independent. If the ACF and PACF of the residuals are significant within the confidence limits, this indicates that there is no significant correlation between the residuals. An alternative method is the Ljung–Box–Pierce (LBQ) test, which is a statistical method of testing for residual autocorrelation. The null hypothesis of the LBQ test is that the residuals are independent. The test statistic is defined in Equation (7):

$$Q = n(n+2) \sum_{k=1}^{m} \frac{r_k^2}{n - k} \tag{7}$$

where m is the number of autocorrelation lags, n is the number of data, and r_k is the sample autocorrelation at lag k. The statistical Q values are compared with the critical value with the degree of freedom at a 5% significance level. If the calculated values are less than the critical value, this means the residuals of the model are in accordance with white noise.

The normality of residuals is verified by histograms and a probability plot of residuals. The residuals are also verified for homoscedasticity, which means there is a constant error variance over all the data. This is verified through a scatterplot of the residuals against predicted values. If the scatterplot exhibits no obvious patterns and the residuals are distributed randomly around zero, this indicates the residuals are homoscedastic.

4. Results and Discussion

In this study, to perform the predictive analysis, we selected seven stations with long-term monitoring data in the study area. The dataset was divided into two periods: training data (from 1967 to 2006) and validation data (from 2007 to 2017). Here, we used the ARIMA model to forecast future drought conditions based on the SPI. The model development steps included model identification, parameter estimation, and diagnostic checking. In this study, we utilized MATLAB to develop the time series model.

4.1. Model Identification

There are two main stages in model identification: (1) confirm whether the data is stationary and (2) utilize the ACF and PACF to determine the general form of the ARIMA model. According to the Kwiatkowski, Phillips, Schmidt, and Shin (KPSS) test, our data was nonstationary, and so it needed to be differenced. After applying the first-order difference, each station satisfied the model development conditions. Because the data at every station exhibited a similar pattern, we simply took the PR1 station as an example. Figure 2 shows that the ACF curve was damping out in a sine wave and the PACF exhibited a significant spike at lag 1, which reflected the AR(1) process. In addition, there were significant spikes in the PACF at lags 12, 24, 36, and 48, which indicated that the data exhibited seasonality with a period of 12. In the ACF, each station had a sine wave pattern, which also indicated that the data was seasonal [13]. Therefore, we chose AR(1) as the nonseasonal part of the ARIMA model, and the peaks at the lags that were multiples of 12 in the PACF indicated a seasonal model, which comprised the SARIMA model. The AIC and SBC criteria were utilized to select the best model, and the results are shown in Table 2. The model with the minimum AIC and SBC was selected as the best model.

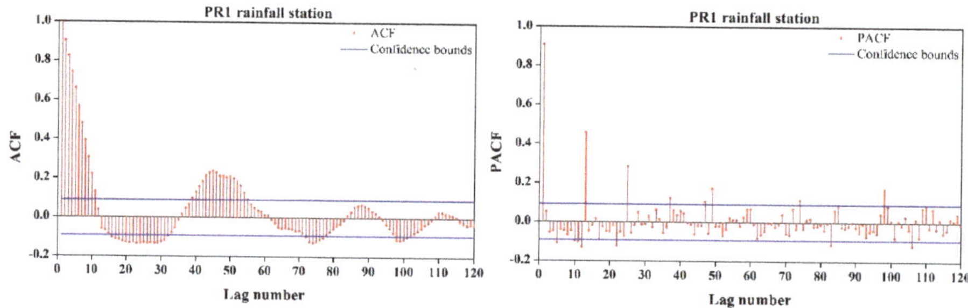

Figure 2. Autocorrelation function (ACF) and partial autocorrelation function (PACF) plots for model selection at the PR1 station in the Pozi River basin.

Table 2. The Akaike information criterion (AIC) and Schwarz–Bayesian criterion (SBC) parameters of each station for selected candidate autoregressive integrated moving average (ARIMA) models.

Station	Model	AIC	SBC	Station	Model	AIC	SBC	Station	Model	AIC	SBC
PR1	SARIMA(1,1,0)(1,0,4)$_{12}$	348.0881	381.4784	YR2	SARIMA(1,1,0)(1,0,3)$_{12}$	288.9087	318.1252	KR3	SARIMA(1,1,0)(2,0,2)$_{12}$	356.0616	385.2781
	SARIMA(1,1,0)(2,0,2)$_{12}$	345.4032	374.6197		SARIMA(1,1,0)(2,0,2)$_{12}$	288.6429	317.8594		SARIMA(1,1,0)(2,0,4)$_{12}$	351.2744	388.8385
	SARIMA(1,1,0)(2,0,3)$_{12}$	347.0453	380.4356		SARIMA(1,1,0)(2,0,3)$_{12}$	277.1375	310.5278		SARIMA(1,1,0)(3,0,2)$_{12}$	355.4204	388.8107
	SARIMA(1,1,0)(2,0,4)$_{12}$	346.5433	384.1074		SARIMA(1,1,0)(2,0,4)$_{12}$	283.4290	320.9931		SARIMA(1,1,0)(3,0,3)$_{12}$	351.1653	388.7294
	SARIMA(1,1,0)(3,0,2)$_{12}$	346.7269	380.1172		SARIMA(1,1,0)(3,0,3)$_{12}$	279.1007	316.6648		SARIMA(1,1,0)(4,0,2)$_{12}$	352.6365	390.2005
	SARIMA(1,1,0)(3,0,4)$_{12}$	314.7873	356.5252		SARIMA(1,1,0)(4,0,4)$_{12}$	271.8073	317.7189		SARIMA(1,1,0)(4,0,3)$_{12}$	352.7802	394.5180
BR1	SARIMA(1,1,0)(1,0,4)$_{12}$	333.3618	366.7521	RR2	SARIMA(1,1,0)(2,0,2)$_{12}$	194.4071	223.6236	LR2	SARIMA(1,1,0)(2,0,3)$_{12}$	354.4606	387.8509
	SARIMA(1,1,0)(2,0,4)$_{12}$	332.7244	370.2284		SARIMA(1,1,0)(2,0,4)$_{12}$	184.9914	222.5554		SARIMA(1,1,0)(2,0,4)$_{12}$	351.3671	388.9312
	SARIMA(1,1,0)(3,0,1)$_{12}$	338.9115	368.1280		SARIMA(1,1,0)(3,0,2)$_{12}$	195.0079	228.3982		SARIMA(1,1,0)(3,0,3)$_{12}$	339.6911	377.2552
	SARIMA(1,1,0)(3,0,2)$_{12}$	331.8391	365.2294		SARIMA(1,1,0)(4,0,0)$_{12}$	190.8300	220.0465		SARIMA(1,1,0)(3,0,4)$_{12}$	338.4294	380.1673
	SARIMA(1,1,0)(3,0,3)$_{12}$	328.9278	366.4919		SARIMA(1,1,0)(4,0,1)$_{12}$	189.1966	222.5869		SARIMA(1,1,0)(4,0,3)$_{12}$	341.4640	383.2019
	SARIMA(1,1,0)(3,0,4)$_{12}$	319.4850	361.2228		SARIMA(1,1,0)(4,0,2)$_{12}$	191.1358	228.6999		SARIMA(1,1,0)(4,0,4)$_{12}$	339.7495	358.6611
GR1	SARIMA(1,1,0)(1,0,3)$_{12}$	238.5829	267.7994								
	SARIMA(1,1,0)(1,0,4)$_{12}$	240.4575	273.8478								
	SARIMA(1,1,0)(2,0,2)$_{12}$	238.4253	267.6418								
	SARIMA(1,1,0)(3,0,2)$_{12}$	239.3507	272.7410								
	SARIMA(1,1,0)(3,0,3)$_{12}$	241.0885	278.6526								
	SARIMA(1,1,0)(4,0,4)$_{12}$	211.4762	257.3878								

4.2. Parameter Estimation

After identifying the order of the model, the parameters must be estimated. Table 3 presents the model parameters, standard errors, *t*-statistics, and *p*-values for the PR1 station in the Pozi River basin. It can be observed that the standard error was reasonably small compared with the model parameters. In addition, most of the *p*-values of the model parameters were less than the alpha level (0.05), which implies that the estimations of the parameters were statistically significant. Therefore, these model parameters should be included in the model.

Table 3. Statistical analysis of the model parameters for the PR1 station in the Pozi River basin.

| Model Parameters | Pozi River Basin: SARIMA(1,1,0)(3,0,4)$_{12}$ | | | |
| | Variables in the Model | | | |
	Value of Parameter	Standard Error	*t*-Statistic	*p*-Value
constant	0.0020	0.0112	0.1765	0.8599
$\varnothing_1$	−0.0903	0.0355	−2.5469	0.0109
Φ_1	−0.2841	0.0286	−9.9461	0.0000
Φ_2	−0.2889	0.0203	−14.2303	0.0000
Φ_3	−0.7950	0.0195	−40.7233	0.0000
Θ_1	−0.4530	0.0366	−12.3870	0.0000
Θ_2	0.1171	0.0360	3.2491	0.0012
Θ_3	0.6851	0.0304	22.5332	0.0000
Θ_4	−0.6389	0.0307	−20.8422	0.0000

$\varnothing_1$ = nonseasonal AR parameter; Φ_1, Φ_2, Φ_3, Φ_4 = seasonal AR parameters; Θ_1, Θ_2, Θ_3, Θ_4 = seasonal MA parameters.

4.3. Diagnostic Checking

After completing the parameter estimation, diagnostic checking was performed. The residuals of a model must be examined to verify that the model is adequate for the time series. The residuals must satisfy the following statistical properties: (1) the residuals are independent of each other; (2) the probability distribution is a normal distribution; and (3) homoscedasticity (constant variance) is satisfied. That is, the residuals must satisfy the requirements of a white noise process. Because each station yielded a similar result, we took the PR1 station as an example.

The independence of the residuals was checked by a correlogram and the LBQ test. Figure 3 shows the RACF and RPACF. The results show that most of the RACF and RPACF values were within the confidence limit, which implies that the residuals did not exhibit a significant correlation with each other. For the second method, the LBQ test was employed in this study to determine whether the residuals are dependent, and the results are presented in Table 4. We can observe that the computed statistics were less than the critical values at each station, which also indicates there was no significant correlation between the residuals, and the residuals from the selected model were in accordance with white noise.

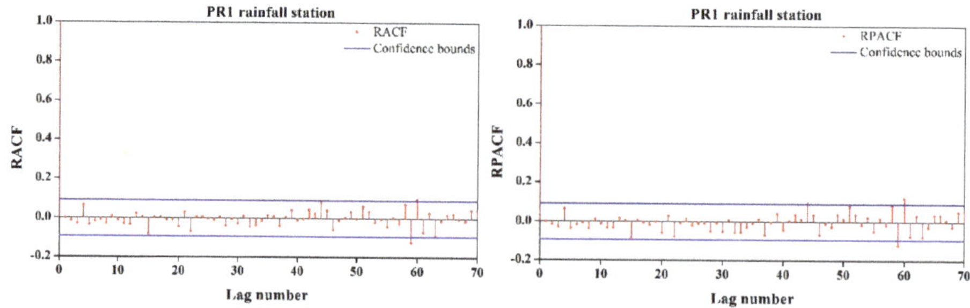

Figure 3. The ACF and PACF of residuals at the PR1 station in the Pozi River basin.

Table 4. Ljung–Box–Pierce (LBQ) statistics for the residuals at each basin.

			LBQ Test				
Station	**PR1**	**BR1**	**GR1**	**YR2**	**RR2**	**KR3**	**LR2**
Test statistic (Q)	55.33	58.41	73.58	53.91	41.43	37.89	48.18
Critical value	89.39	89.39	89.39	89.39	89.39	89.39	89.39

Degrees of freedom = 69; significance level = 0.05.

Figure 4 depicts the histogram and normal probability plot of the residuals at the PR1 station in the Pozi River basin. The histograms show that the residuals were roughly centered on zero and were more or less normally distributed [13,33]. The normal probability plot of the residuals indicates that the residuals lay on a diagonal line, which represents the normal probability for residuals in each basin [13,34]. Therefore, both methods provided evidence of the normality of the residuals.

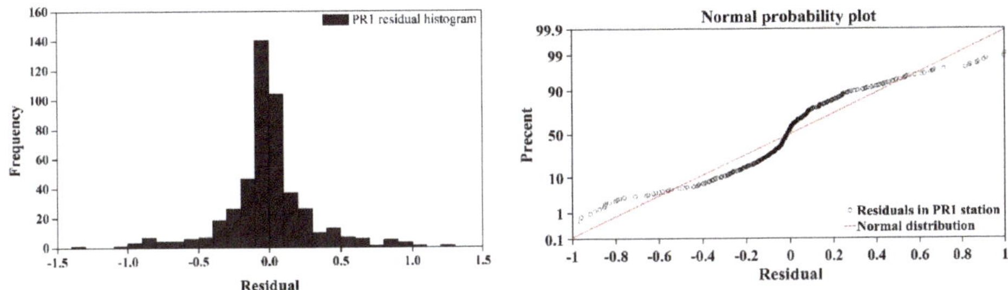

Figure 4. The histogram (**left column**) and normal probability plot (**right column**) of residuals at the PR1 station in the Pozi River basin.

To determine whether the ability of the model to predict variable values is consistent, it is important to verify the homoscedasticity of the residuals [35]. The homoscedasticity of the residuals was checked by the scatterplot of the residuals against predicted values. The results are shown in Figure 5. The plot exhibited no pattern, and the residuals were randomly scattered. That is, the residuals were evenly distributed around a zero mean, which implies that the model was well fitted.

According to the above check, the residuals of the model were uncorrelated, had constant variance, and were normally distributed, and the statistical properties of the residuals were compliant with white noise. Thus, we confirmed that the selected model is adequate for the corresponding SPI time series at each station.

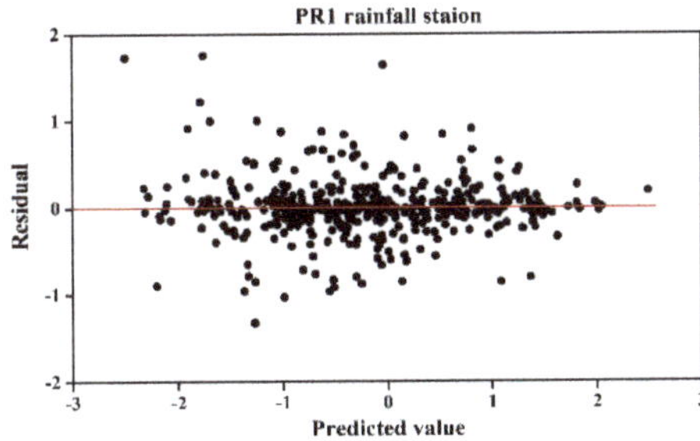

Figure 5. The scatterplot of the residuals against predicted values at the PR1 station.

4.4. Model Validation

Here, we utilized data from 2007 to 2017 for model validation. Figure 6 presents a comparison of the observed data with predicted values at each basin using the best SARIMA model. The predicted value yielded a similar pattern to the observed data, and the performance measures are presented in Table 5. In general, the higher the R^2 value, the better the performance of the model. According to our results, the R^2 values in each station were greater than 0.8, and the root-mean-square error (RMSE) and mean absolute error (MAE) were also sufficiently low. Therefore, the SARIMA model used to predict drought index in this study is reasonably precise.

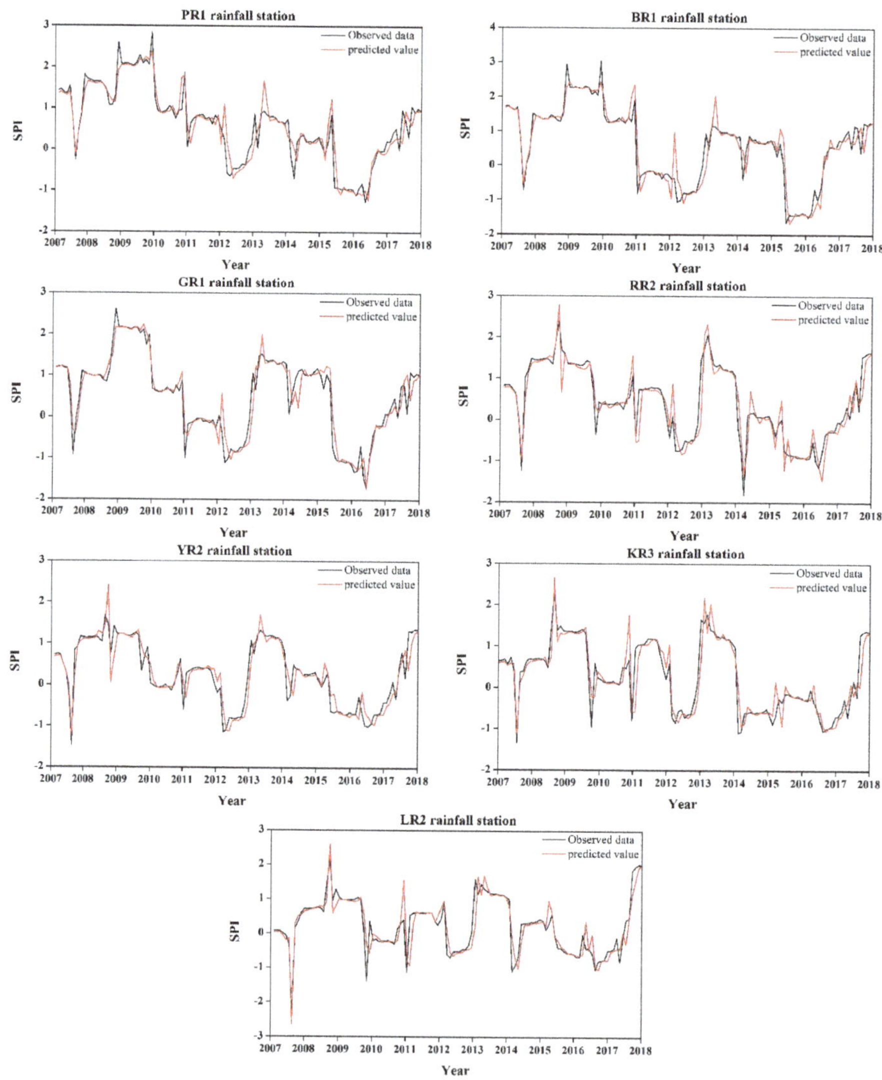

Figure 6. Comparison of observed data with predicted values using the best seasonal ARIMA (SARIMA) model at each basin.

Table 5. Performance measures for the selected model for the observed data and predicted values.

Station	Model	Performance Measures		
		R_2	RMSE	MAE
PR1	ARIMA$(1,1,0)(3,0,4)_{12}$	0.8775	0.3220	0.2159
BR1	ARIMA$(1,1,0)(3,0,4)_{12}$	0.8869	0.3628	0.2291
GR1	ARIMA$(1,1,0)(4,0,4)_{12}$	0.8938	0.3241	0.2120
YR2	ARIMA$(1,1,0)(2,0,3)_{12}$	0.8445	0.3123	0.2060
RR2	ARIMA$(1,1,0)(2,0,4)_{12}$	0.8602	0.3482	0.2239
KR3	ARIMA$(1,1,0)(3,0,3)_{12}$	0.8337	0.3699	0.2339
LR2	ARIMA$(1,1,0)(3,0,3)_{12}$	0.8213	0.3714	0.2281

Abbreviations: RMSE—root-mean-square error, MAE—mean absolute error.

4.5. Forecasting

The reason that drought forecasting is necessary for water resource management and planning is that drought events can then be diagnosed in advance, so that experts can take precautions. In this study, we employed the seasonal ARIMA model to forecast the drought condition in the next four years, from 2019 to 2022. The results at each basin are depicted in Figure 7. It was observed that the predicted values showed stochastic change for each SPI time series. According to the analytical results, the lowest SPI values were concentrated in the summer, which means that this region may be affected by drought in the summer in the future. In general, the main source of rainfall in Taiwan during the summer is typhoons. Typhoons can introduce abundant water resources. However, if the number of typhoons is small or the typhoons do not directly affect Taiwan, this sometimes results in a water shortage in the summer. In 2018, the number of typhoons generated between June and September was as high as 22. This is the year in which the most typhoons occurred in the past 10 years. However, only two typhoons affected Taiwan, which may have led to insufficient summer rainfall in Taiwan [36]. In addition, the Pacific high is the main climate factor during the summer in Taiwan, and this is becoming stronger owing to climate change [37], making Taiwan's climate hotter and with less precipitation. The Pacific high also guides the paths of typhoons. If the western edge of the Pacific high extends westward to the Asian continent, then a typhoon may pass through the Philippines instead of Taiwan. This is one of the reasons why there were fewer typhoons in 2018 in Taiwan. In recent years, the Pacific high has exhibited a tendency to move increasingly westward [38]. Therefore, the summer climate of Taiwan may change in the future under climate change. However, the details remain to be discussed in future research. In this study, the ARIMA model was employed to forecast drought events in the future. Unlike previous studies, we used a stochastic rather than deterministic method to describe the forecasting results. According to historical drought events in Taiwan, drought events occur approximately every three to four years. There was a water shortage situation in 2017 in Taiwan, and 2020 may be a drier year, which is consistent with the results we found in this study. A statistical method involves establishing suitable models to characterize climate factors such as precipitation. Stationarity is generally assumed to exist between predictor and historical data, but this is not always true. Thus, the uncertainty of the model may be higher. In addition, there are many factors that affect environmental changes, and this may present a challenge under this situation. Further studies should consider climate variability in the model for drought forecasting.

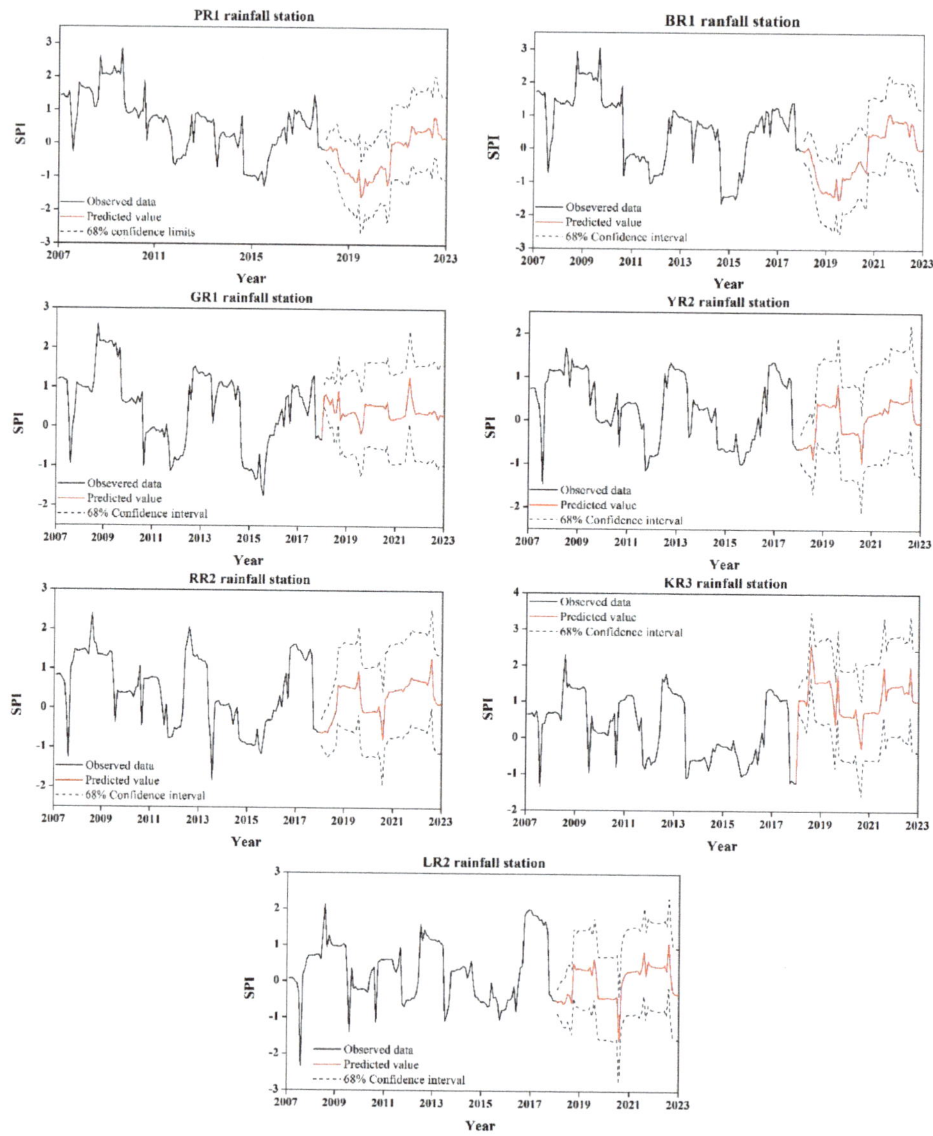

Figure 7. Drought forecasting for the period 2018–2022 using the seasonal ARIMA model at each basin.

5. Conclusions

In this study, the ARIMA model was employed as a drought forecasting tool in southern Taiwan. We used data from 1967 to 2006 to train the model. The model development included three steps: model identification, parameter estimation, and diagnostic checking. In the model identification step, we selected the general form of the model and chose the model with the minimum AIC and SBC as the best fit. In the parameter estimation step, we utilized several statistics to determine whether the parameters we estimated were significant. The results showed that most of the model parameters had p-values below the alpha level (0.05). The diagnostic checking step indicated that the statistical properties of the model residuals were compliant with white noise, including being uncorrelated with a constant variance and normal distribution. Then, the data from 2007 to 2017 were used to validate the model. The results showed that there was a high coefficient of determination (R^2) at each station (all over 0.80) and low values for the RMSE and MAE, which implies that the model is adequately precise at each basin. Finally, we used the ARIMA model to forecast the future drought conditions from 2019 to 2022. The forecasting results demonstrate that the SPI value is relatively low in the summer of 2020, which implies that there may be a water shortage in southern Taiwan. This phenomenon may be related to climate change, which leads to an enhancement of the Pacific high extending westward, thus affecting the paths of typhoons. In addition, the Pacific high dominates the summer climate in Taiwan, and if its intensity continues to increase, this will reduce precipitation in Taiwan in the future. However, the detailed evolution mechanism still remains to be discussed in the future.

The stochastic models selected for forecasting the SPI time series provided information on precipitation in southern Taiwan. This is a powerful tool, which can also be used to describe the hydrological time series. The ARIMA model used in this study, based on the SPI, can be applied to forecast drought impacts, playing a vital role in mitigating drought in water resource systems. However, the corresponding natural phenomena are complicated, owing to the influences of many factors. Stochastic models do not consider physical processes, and so it is difficult to understand the physical mechanisms of climate change. In addition, the assumption of the model may lead to higher uncertainty. Nevertheless, the model can be used to predict future trends and serve as a variable for other physical models in further studies.

Author Contributions: H.-F.Y. conceived of the subject of the article, performed the literature review, and contributed to the writing of the paper; H.-L.H. participated in data processing and the elaboration of the statistical analysis and figures.

Funding: This research received no external funding.

Acknowledgments: The authors are grateful for the support from the Headquarters of University Advancement at the National Cheng Kung University, sponsored by the Ministry of Education, Taiwan, ROC.

Conflicts of Interest: The authors declare no conflict of interest.

References

1. Dai, A. Increasing drought under global warming in observations and models. *Nat. Clim. Chang.* **2013**, *3*, 52–58. [CrossRef]
2. Habibi, B.; Meddi, M.; Torfs, P.J.; Remaoun, M.; Van Lanen, H.A. Characterisation and prediction of meteorological drought using stochastic models in the semi-arid Chéliff–Zahrez basin (Algeria). *J. Hydrol. Reg. Stud.* **2018**, *16*, 15–31. [CrossRef]
3. Tsakiris, G. Drought Risk Assessment and Management. *Water Resour. Manag.* **2017**, *31*, 3083–3095. [CrossRef]
4. Belayneh, A.; Adamowski, J.; Khalil, B.; Ozga-Zielinski, B. Long-term SPI drought forecasting in the Awash River Basin in Ethiopia using wavelet neural network and wavelet support vector regression models. *J. Hydrol.* **2014**, *508*, 418–429. [CrossRef]
5. Bordi, I.; Sutera, A. Drought monitoring and forecasting at large scale. In *Methods and Tools for Drought Analysis and Management 2007*; Springer: Dordrecht, The Netherlands, 2017; pp. 3–27.
6. Mishra, A.K.; Singh, V.P. Drought modeling—A review. *J. Hydrol.* **2011**, *403*, 157–175. [CrossRef]
7. Mossad, A.; Alazba, A.A. Drought forecasting using stochastic models in a hyper-arid climate. *Atmosphere* **2015**, *6*, 410–430. [CrossRef]
8. Panu, U.S.; Sharma, T.C. Challenges in drought research: Some perspectives and future directions. *Hydrol. Sci. J.* **2002**, *47*, S19–S30. [CrossRef]
9. Li, J.; Zhou, S.; Hu, R. Hydrological drought class transition using SPI and SRI time series by loglinear regression. *Water Resour. Manag.* **2016**, *30*, 669–684. [CrossRef]
10. Park, S.; Im, J.; Jang, E.; Rhee, J. Drought assessment and monitoring through blending of multi-sensor indices using machine learning approaches for different climate regions. *Agric. For. Meteorol.* **2016**, *216*, 157–169. [CrossRef]

11. Durdu Ömer, F. Application of linear stochastic models for drought forecasting in the Büyük Menderes river basin, western Turkey. *Stoch. Environ. Res. Risk Assess.* **2010**, *24*, 1145–1162. [CrossRef]

12. Bazrafshan, O.; Salajegheh, A.; Bazrafshan, J.; Mahdavi, M.; Fatehi Maraj, A. Hydrological drought forecasting using ARIMA models (Case study: Karkheh Basin). *Ecopersia* **2015**, *3*, 1099–1117.

13. Mahmud, I.; Bari, S.H.; Rahman, M.T.U. Monthly rainfall forecast of Bangladesh using autoregressive integrated moving average method. *Environ. Eng. Res.* **2016**, *22*, 162–168. [CrossRef]

14. Karthika, K.; Thirunavukkarasu, V.; Karthika, M. Forecasting of meteorological drought using ARIMA model. *Indian J. Agric. Res.* **2017**, *51*, 103–111. [CrossRef]

15. Rahmat, S.N.; Jayasuriya, N.; Bhuiyan, M.A. Short-term droughts forecast using Markov chain model in Victoria, Australia. *Theor. Appl. Climatol.* **2017**, *129*, 445–457. [CrossRef]

16. Agboola, A.H.; Gabriel, A.J.; Aliyu, E.O.; Alese, B.K. Development of a fuzzy logic based rainfall prediction model. *Int. J. Eng. Technol.* **2013**, *3*, 427–435.

17. Jalalkamali, A.; Moradi, M.; Moradi, N. Application of several artificial intelligence models and ARIMAX model for forecasting drought using the Standardized Precipitation Index. *Int. J. Environ. Sci. Technol.* **2015**, *12*, 1201–1210. [CrossRef]

18. Borji, M.; Malekian, A.; Salajegheh, A.; Ghadimi, M. Multi-time-scale analysis of hydrological drought forecasting using support vector regression (SVR) and artificial neural networks (ANN). *Arab. J. Geosci.* **2016**, *9*, 725. [CrossRef]

19. Kousari, M.R.; Hosseini, M.E.; Ahani, H.; Hakimelahi, H. Introducing an operational method to forecast long-term regional drought based on the application of artificial intelligence capabilities. *Theor. Appl. Climatol.* **2017**, *127*, 361–380. [CrossRef]

20. Sheffield, J.; Wood, E.; Chaney, N.; Guan, K.; Sadri, S.; Yuan, X.; Olang, L.; Amani, A.; Ali, A.; DeMuth, S.; et al. A drought monitoring and forecasting system for sub-Sahara African water resources and food security. *Bull. Am. Meteorol. Soc.* **2014**, *95*, 861–882. [CrossRef]

21. Dehghani, M.; Saghafian, B.; Rivaz, F.; Khodadadi, A. Evaluation of dynamic regression and artificial neural networks models for real-time hydrological drought forecasting. *Arab. J. Geosci.* **2017**, *10*, 266. [CrossRef]

22. Alsharif, M.H.; Younes, M.K.; Kim, J. Time series ARIMA model for prediction of daily and monthly average global solar radiation: The case study of Seoul, South Korea. *Symmetry* **2019**, *11*, 240. [CrossRef]

23. Mishra, A.K.; Desai, V.R. Drought forecasting using stochastic models. *Stoch. Environ. Res. Risk Assess.* **2005**, *19*, 326–339. [CrossRef]

24. McKee, T.B.; Doesken, N.J.; Kleist, J. The relationship of drought frequency and duration to time scales. In Proceedings of the Eighth Conference on Applied Climatology, Boston, MA, USA, 17–22 January 1993; Volume 17, pp. 179–183.

25. WMO. *Standardized Precipitation Index User Guide*; Svoboda, M., Hayes, M., Wood, D.A., Eds.; WMO: Geneva, Switzerland, 2012.

26. Bari, S.H.; Rahman, M.T.; Hussain, M.M.; Ray, S. Forecasting monthly precipitation in Sylhet city using ARIMA model. *Civ. Environ. Res.* **2015**, *7*, 69–77.

27. Box, G.E.P.; Jenkins, G.M. *Time Series Analysis: Forecasting and Control*; Holden-Day: San Francisco, CA, USA, 1976.

28. Box, G.E.P.; Jenkins, G.M.; Reinsel, G.C. *Time Series Analysis, Forecasting and Control*; Prentice Hall: Englewood Cliffs, NJ, USA, 1994.

29. Modarres, R. Streamflow drought time series forecasting. *Stoch. Environ. Res. Risk Assess.* **2007**, *21*, 223–233. [CrossRef]

30. Brockwell, P.J.; Davis, R.A. *Introduction to Time Series and Forecasting*; Springer Science and Business Media LLC: Berlin/Heidelberg, Germany, 2016.

31. Akaike, H. A New look at the statistical model identification. *Funct. Shape Data Anal.* **1974**, *19*, 215–222.

32. Schwarz, G. Estimating the dimension of a model. *Ann. Stat.* **1978**, *6*, 461–464. [CrossRef]

33. Rahman, M.A.; Yunsheng, L.; Sultana, N. Analysis and prediction of rainfall trends over Bangladesh using Mann–Kendall, Spearman's rho tests and ARIMA model. *Meteorol. Atmos. Phys.* **2017**, *129*, 409–424. [CrossRef]

34. Widowati; Putro, S.P.; Koshio, S.; Oktaferdian, V. Implementation of ARIMA model to asses seasonal variability macrobenthic assemblages. *Aquat. Procedia* **2016**, *7*, 277–284. [CrossRef]

35. Huang, Y.F.; Mirzaei, M.; Yap, W.K. Flood analysis in Langat river basin using stochastic model. *Int. J. GEOMATE* **2016**, *11*, 2796–2803.
36. *Central Weather Bureau, Monthly Report on Climate System: Typhoon Climate Analysis*; Central Weather Bureau, Ministry of Transportation and Communications: Taipei, Taiwan, 2019.
37. Li, W.; Li, L.; Ting, M.; Liu, Y. Intensification of Northern Hemisphere subtropical highs in a warming climate. *Nat. Geosci.* **2012**, *5*, 830–834. [CrossRef]
38. Taiwan Climate Change Projection and Information Platform (TCCIP). *Taiwan Climate Change Science Report 2017—Physical Phenomena and Mechanisms*; TCCIP: Taipei, Taiwan, 2017.

Article

Attribution Analysis of Hydrological Drought Risk Under Climate Change and Human Activities: A Case Study on Kuye River Basin in China

Ming Zhang *, Jinpeng Wang and Runjuan Zhou

School of Civil Engineering, Anhui Polytechnic University, Wuhu 241000, China;
2170430104@stu.ahpu.edu.cn (J.W.); rjzhou@126.com (R.Z.)
* Correspondence: ahzhming@163.com; Tel.: +86-553-287-1178

Received: 8 August 2019; Accepted: 18 September 2019; Published: 20 September 2019

Abstract: This study conducted quantitative diagnosis on the impact of climate change and human activities on drought risk. Taking the Kuye river basin (KRB) in China as the research area, we used variation point diagnosis, simulation of precipitation and runoff, drought risk assessment, and attribution quantification. The results show that: (1) the annual runoff sequence of KRB changed significantly after 1979, which was consistent with the introduction of large-scale coal mining; (2) under the same drought recurrence period, the drought duration and severity in the human activity stage were significantly worse than in the natural and simulation stages, indicating that human activities changed the drought risk in this area; and (3) human activities had little impact on drought severity in the short duration and low recurrence period, but had a greater impact in the long duration and high recurrence period. These results provide scientific guidance for the management, prevention, and resistance of drought; and guarantee sustainable economic and social development in the KRB.

Keywords: drought risk; climate change; human activities; quantitative attribution; artificial neural network

---

1. Introduction

Drought, an abnormal water shortage caused by imbalance in the water cycle, is an extreme event in the water cycle. Many definitions of drought exist, including meteorological drought, agricultural drought, socio-economic drought, and hydrologic drought [1,2]. The latter is defined as a deficit in surface water or groundwater, including a reduction in water supply for drinking water, irrigation, industrial needs, and hydropower production, causing the death of fish and hampering navigation [1].

The change in the hydrological drought risk of watersheds is the result of environmental change, mainly modification in global natural conditions, such as climate change and human activities, which can have strong local impacts and contribute to global effects. Influenced by climate change marked by global warming, and driven by high-impact human activities, global change has become obvious over the past 100 years. The water cycle process has changed to different degrees, reflected in the increasing risk of natural disasters represented by flood and drought disasters [3–5]. The occurrence of drought events is worsening, with higher frequency, greater severity, and wider scope [6]. Serious and extreme drought events are increasing, as is the destruction caused by drought disasters [7], posing a serious threat to human living environments, food security, social stability, and sustainable economic development. With climate warming, population growth, and economic development, the frequency of drought in Northwest China is among the highest in the country. Large-scale severe drought has been frequently recorded, resulting in increasingly serious losses and impacts on the environment. Therefore, drought is known to seriously threaten sustainable economic and social development [8].

On analyzing several research works, we found that drought is affected by climatic factors (natural climate and anthropogenic climate), environmental resource factors, topographic and geomorphic factors, water resource conditions, social and economic factors, population growth factors, coal mining, water conservancy facilities, and other land use factors [7,9–13]. Strzepek et al. [14] and Kirono et al. [15] used global climate models to study the factors affecting drought in Australia and most parts of the United States, respectively. The results show that the emission of aerosols will increase the area affected by drought and escalate the frequency of drought. Wang et al., based on 35 climate models and two emissions scenarios (RCP4.5 and RCP8.5), applied the coupled model inter-comparison project phase 5 (CMIP5) model to forecast China's future drought situation; the results showed that China's drought situation may worsen with temperature increases [16]. Based on the measured meteorological data of 101 meteorological stations in the Yellow River basin and the precipitation and temperature data of six models from 1961 to 2099 under the three emission scenarios of CMIP5, Yang et al. (2018) assessed the drought in the Yellow River basin, and the results show that the drought conditions in the early 21st century under the three emission scenarios were all severe, relative to the baseline period [17]. According to Williams et al.'s analysis of drought in California, the main driving factor is precipitation. Although natural variation is dominant, the climate warming caused by human activities intensifies the probability of drought occurrence in California, and with a continuous increase in temperature, water will evaporate from plants and soil into the air, leading to the aggravation of drought severity in the state [18].

Therefore, calculating the proportion of the contribution of climate change and human activities to the risk of drought is important to guide the work of local governments in preventing and combating drought. However, little research has been conducted on the issue. For example, Jiang et al. used empirical parameter statistical analysis to separate and quantitatively evaluate the factors impacting flood disaster area in Xinjiang for more than 50 years; the results show that flood disasters in Xinjiang were mainly affected by human activities and precipitation anomalies [19]. This research idea can be applied to the quantitative attribution of drought risk change. Al-faraj et al. showed that the construction of reservoirs in upstream regions in the Diyala River basin in central Asia significantly changed the hydrological situation in the downstream region, aggravating the hydrological drought [20]. Zhang et al. studied the drought risk in the Huaihe River basin, showing that the construction and operation of reservoirs can reduce the occurrence frequency of hydrological drought in the downstream region and extend the duration of drought [21]. Luan et al. drew a drought scale map and found that since the 1960s, drought events in the Fen and Wei plains have been increasing, but the occurrence frequency of extreme drought and excessive drought has been significantly reduced, which is related to the introduction of several water conservancy projects during this period, which reduced the harm level of drought events to some extent [22]. Based on the monthly runoff data of Boluo station in the Dongjiang River basin, Tu et al. constructed multivariate joint distribution of hydrological and drought indicators using the meta-Gaussian Copula function, and quantitatively assessed the impact of reservoir runoff regulation on hydrological and drought multivariate joint characteristics. The results showed that the regulation of reservoir runoff alleviated the hydrological drought in the Dongjiang River basin, but the multi-variable joint transcendence period of the same group of drought indicators significantly increased under the influence of the reservoir [23].

However, most studies on the impact of climate change and human activities on drought risk are qualitative or semi-quantitative, and very few have focused on drought risk change. With the increasingly severe impacts of climate change and human activities on hydrology and water resources in river basins, non-quantitative research does not help in understanding the internal mechanism of hydrology [24]. Determining how to quantitatively distinguish the impacts of climate change and human activities on the hydrological cycle and the water resource formation process within a region has become a focus of academic research [25,26]. The lack of hydrological, soil, and vegetation data in the region cannot meet the requirements of climate and hydrological models. Rainfall and runoff data, which are easy to obtain, directly reflect change in water resources. Therefore, we propose a variation attribution analysis method based on the relationship between precipitation and runoff. On the basis of identifying and dividing the reference period, this method starts from the runoff, showing the change in hydrological and drought risk and the precipitation reflecting climate change, and simulates a runoff series in the variation period only subject to climate change. Then, the monthly runoff of the three stages (natural stage, human activity stage, and simulation stage) is obtained, and the recurrence period, duration, and severity of drought are calculated according to the runoff of the three stages, such that the contribution rate of various factors affecting the variation of drought risk can be quantitatively calculated. A case study of the Kuye river basin (KRB) in China was conducted to verify the effectiveness of the method.

The contributions of this method include: (1) coupling the precipitation and runoff simulation and hydrological sequence variation points of diagnosis and methods, such as hydrologic drought risk assessment; (2) revealing the factors driving drought risk in terms of climate change and human activities on drought duration, drought severity, and return period of drought to guide disaster prevention and mitigation practice; and (3) introducing an idea to study the variation in drought risk, which can be used for attribution diagnosis of meteorological drought, agricultural drought, and social and economic drought.

The following sections are arranged as follows. Section 2 introduces the framework of our method in detail, including the ordered clustering method (OCM) for the diagnosis of variation points, the artificial neural network (ANN) method to simulate the relationship between precipitation and runoff, the Copula method for drought risk assessment, and the attribution diagnosis method for hydrological factors. Section 3 introduces the hydrology and geography of the KRB. Section 4 applies the above method framework and takes the KRB as an example to obtain variation point diagnosis results, ANN simulation results, three-stage Copula drought risk assessment results, results of climate change, and the human activity contribution ratio. Section 5 summarizes our conclusions.

2. Methodology

2.1. Ordered Clustering Method (OCM)

As global climate change and human activities increase, the statistical scale of hydrological elements have changed, and the hydrological series, including precipitation and runoff, which are closely related to human production and life, have lost their original consistency. Scholars have conducted extensive and effective work in the diagnosis of water resource variation [26,27]. Xie et al. proposed an approach to evaluate the significance of abrupt changes in time series at five levels: no, weak, moderate, strong, and dramatic change; and stated that human activities contributed much more than climate change to the abrupt changes in the corresponding surface water resources amount [26]. Song et al. studied the changes in the frequency of precipitation extremes and their contributions to total precipitation at different extreme percentile thresholds. They found that extreme precipitation events with high-percentile thresholds have more spatial and temporal variability than those with low-percentile thresholds [27]. The affected series and the original series (natural series) of data can be regarded as two different classes. Therefore, OCM can be used to deduce mutation points between the natural series and the affected series.

OCM is essentially used to deduce the optimal segmentation point to minimize the sum of squares of deviations among similar classes, when the sum of squares of deviations between classes is relatively large [28,29]. If the hydrological series has two distinct stages, the time series changes of the sum of squares of the total deviations show a single valley bottom, and the year corresponding to the valley bottom is the optimal series mutation year. The optimal segmentation method is as follows:

$$V_\tau = \sum_{i=1}^{\tau} (x_i - \overline{x_\tau})^2 \tag{1}$$

$$V_{n-\tau} = \sum_{i=\tau+1}^{n} \left(x_i - \overline{x_{n-\tau}}\right)^2 \tag{2}$$

where n is the length of the series, and τ is the possible variation position; $\overline{x_\tau}$ and $\overline{x_{n-\tau}}$ are the mean values of hydrological series before and after the variation points, respectively; V_τ and $V_{n-\tau}$ are the sum of deviation squares of each stage.

The total sum of squares of deviations can be calculated as:

$$S_n(\tau) = V_\tau + V_{n-\tau} \tag{3}$$

when the $S_n(\tau)$ is a minimum value and the corresponding point τ is the optimal point of division.

2.2. ANN Method to Simulate the Relationship Between Precipitation and Runoff

Artificial neural network (ANN) has attracted attention due to its advantages in solving highly nonlinear problems [30]. The back-propagation (BP) algorithm, which was used to train the ANN models, is the most popular training algorithm for Multilayer Perception (MLP)—ANN models, which are the most applied ANN types in the field of hydrology [31]. Experts have extensively applied the BP ANN to the study of precipitation and runoff simulation over the years, perfecting its theory and technology [32–35].

When we used the BP network to simulate the relationship between precipitation and runoff in KRB, we mainly used the precipitation of the first six months to simulate the monthly runoff, so the 6-10-1 network structure was adopted for simulation. When choosing the month of precipitation data, we considered that the model should reflect the change in soil water storage in the basin, and the number of months should not be too small; to make the model more concise, the input data should not be excessive, so we chose precipitation prediction runoff of the first six months. The calculation process was mainly simulated using the Visual C++ 2010 (Microsoft Corporation, Redmond, DC, USA) programming platform.

The Nash–Sutcliffe efficiency (NSE) coefficient, commonly used in the field of hydrological simulation, was adopted as the evaluation tool for the simulation effect [36]:

$$NSE = 1 - \frac{\sum_{i=1}^{N}\left(Q_{oi} - Q_{si}\right)^2}{\sum_{i=1}^{N}\left(Q_{oi} - \overline{Q_o}\right)^2} \tag{4}$$

where Q_{oi} is the observed runoff, Q_{si} is the simulated runoff, and $\overline{Q_o}$ is the total average of the observed runoff. *NSE* has a range of $-\infty$ to 1. An *NSE* value close to 1 indicates that the model is of good quality and high reliability. An *NSE* close to or less than 0 indicates that the output value of the model is too large from the observed value, and the cause of the problem needs to be found.

2.3. Drought Risk Assessment Method

The model includes the following four steps.

2.3.1. Step 1: Identification of Drought Processes

The run-length theory [1,37] is generally adopted to set the threshold value of the hydrological series. In this paper, this hydrological series is the monthly runoff depth, which represents the river flow discharge. The runoff falling below a certain threshold is recognized as a drought event. The threshold value can usually be determined by the percentage of the historical mean of runoff in a certain period, such as −20% and −40% [37]. According to the run-length theory and the set threshold, the drought duration and corresponding drought severity can be identified. A schematic diagram identifying the drought process is shown in Figure 1, where a, b, c and d represent 4 drought events. We then obtained the drought duration (D) and drought severity (S) of each drought event.

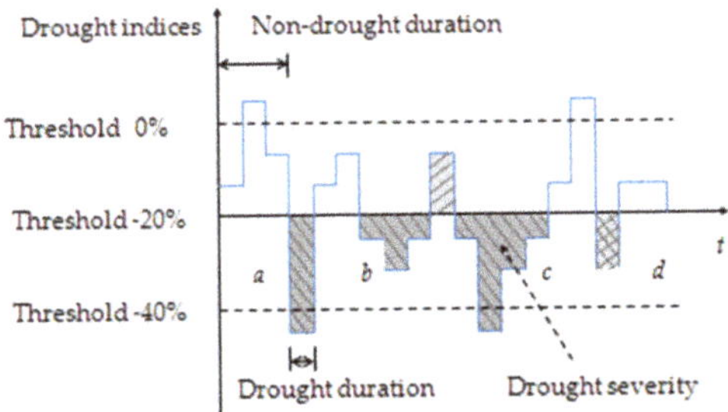

Figure 1. Schematic diagram of drought identification.

2.3.2. Step 2: Determination of Marginal Distribution of Drought Characteristics

The marginal distribution of drought duration and severity are the basis for determining the joint distribution of drought risk. Generally, γ distribution and its deformation forms, such as exponential distribution and Pearson type III distribution, can be adopted [27,37,38]. Considering the generality of the marginal distribution calculation method, we uniformly adopted γ distribution to determine the marginal distribution of drought characteristics. Then, we obtained the probability density function of drought duration $u(d)$ and drought severity $v(s)$ and the distributed function $F_D(d)$ and $F_S(s)$.

2.3.3. Step 3: Determination of Joint Distribution of Drought Risk

Copula function is widely used to measure the joint distribution of drought risk. Various forms of the Copula function exist, among which Gumbel–Hougaard (GH) Copula in Archimedean Copula is the most commonly used, including in the field of hydrology [37,38]. The function form of GH Copula is as follows:

$$F_{D,S}(d,s) = C(u,v) = \exp\left\{-\left[(-\ln u)^\theta + (-\ln v)^\theta\right]^{\frac{1}{\theta}}\right\} \tag{5}$$

where the parameter θ can be obtained through the Kendall correlation coefficient between drought duration and drought severity. Therefore, we also used the GH function to determine the joint distribution of drought duration and drought severity.

2.3.4. Step 4: Recurrence Period of the Joint Distribution of Hydrological Drought Risk

According to joint distribution of the above Copula functions in Equation (3), when drought duration $D > d$ and drought severity $S > s$, the drought recurrence period $T(d,s)$ is [39]:

$$T(d,s) = \frac{E(L)}{P(D > d \cap S > s)} = \frac{E(L)}{1 - F_D(d) - F_S(s) + F_{D,S}(d,s)} \tag{6}$$

where $E(L)$ is the expectation of the interval between two drought events, equal to the sum of the average of drought duration and non-drought duration.

2.4. Quantitative Identification of Drought Risk Attribution

The precipitation and runoff simulation approach was adopted to simulate the runoff sequence without the influence of human activities, and then the hydrological drought recurrence period of each stage can be calculated according to the GH Copula function to quantitatively identify the attribution of drought risk change.

We suppose that the variation points are divided into three stages: natural stage 1 (observation), human activities stage 2 (observation), and ANN-simulated stage 2 (simulation). Under a certain drought recurrence period T and drought duration D, the first stage has drought severity S_1 and runoff R_1, and the second observation stage has severity S_2 and runoff R_2, as shown in Figure 2. According to the ANN simulation in Section 2.2, the functional relationship between R_1 and P_1 was established, $R_1 = F(P_1)$, and, combined with the hydrological drought risk Copula calculation function, the runoff R_{2s} obtained after the simulation can be used to calculate the drought severity under each recurrence period and drought duration.

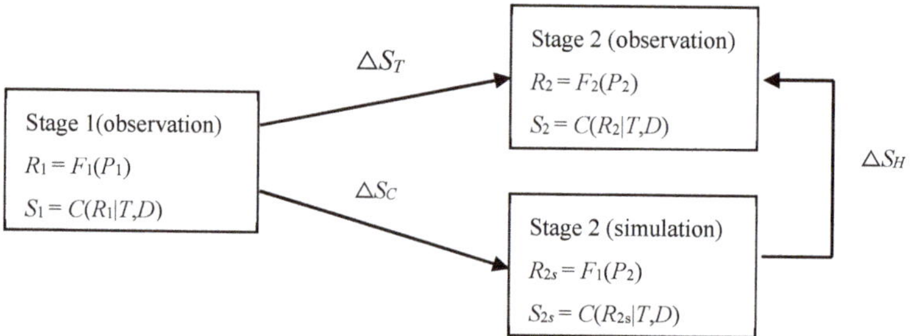

Figure 2. Schematic diagram of quantitative identification of drought risk attribution.

For the first stage, we have $S_1 = C(R_1|T, D)$, where the F function constitutes the factors of the underlying surface. If human activities have no influence, we think that this F function is also applicable to the second stage. However, according to the calculation of R_{2s} and Copula function, a large difference ΔS_H exists between S_{2s} and the observed S_2. This indicates that the underlying surface of the second stage has undergone significant changes under the influence of human activities. As the underlying surface F adopted by S_{2s} in the simulation calculation is the same as that of the first stage, the difference in the value between S_{2s} and S_1, ΔS_C, reflects the influence of climate change.

In Figure 2, the difference between S_1 and S_2, ΔS_T, reflects the total change in drought severity. It has two parts: one caused by human activities, and the other by climate change.

The following equations outline the method of analyzing the impact of environmental change on runoff:

$$\Delta S_T = S_2 - S_1 \tag{7}$$

$$\Delta S_H = S_2 - S_{2s} \tag{8}$$

$$\Delta S_C = S_{2s} - S_1 \tag{9}$$

$$\eta_H = \frac{\Delta S_H}{\Delta S_T} \times 100\% \tag{10}$$

$$\eta_C = \frac{\Delta S_C}{\Delta S_T} \times 100\% \tag{11}$$

where η_H and η_C are the percentages of impacts of human activities and climate change on the contribution to hydrological drought severity, respectively.

3. Site Description

The KRB is located between 109°28′ and 110°52′ E, and 38°23′ and 39°52′ N, with an area of about 8706 km² and total length of 242 km. Elevation of the KRB is about 800–1300 m, with a river source elevation of 1498.7 m and estuary elevation of 740.6 m, which is a total drop of 758.1 m with average river slope of 2.55‰. The terrain in this region is high in the northwest and low in the southeast, and the depth of gullies is more than 150 m. The basin has a warm temperate semi-arid continental monsoon climate: dry and windy in spring, hot and rainy in summer, rainy and cold in autumn, and dry and cold in winter. The annual average temperature is 8.9 °C, ranging from −28.1 °C to 38.9 °C, with snow cover in winter. The average annual precipitation is 375–450 mm, increasing from northwest to southeast, mainly occurring from July to October, accounting for more than 70% of the annual precipitation.

We collected monthly precipitation data from 17 rain-measuring stations in the KRB and monthly runoff data from the Wenjiachuan hydrological station from 1954 to 2005. The Wenjiachuan station is the most downstream control station along the Kuye River, which flows into the Yellow River in Shamao Village, Shaanxi Province, China [40]. The locations of the 17 rain-measuring stations and the Wenjiachuan hydrology station are shown in Figure 3.

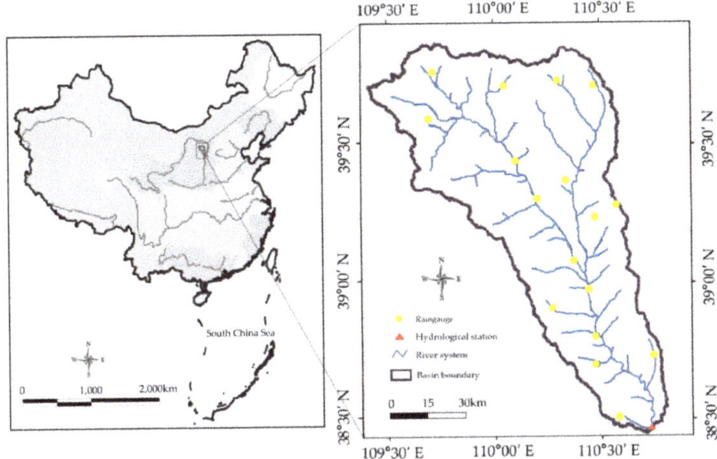

Figure 3. River system and location of the Kuye river basin (KRB), China.

The conditions of the underlying surface of the KRB are complex. The soil in the watershed is loose and barren, and soil erosion is serious due to the lack of vegetation on the surface. The KRB is rich in high-quality coal resources, and the famous Shenfu-Dongsheng coal field runs through the middle of the basin. In the 1980s, with the focus shifting from coal resources development in China to the fragile ecological environment in the west, large-scale coal mining began in the KRB [41,42].

4. Results and Discussion

4.1. Variation Point Diagnosis

To qualitatively understand the changes in precipitation and runoff in the KRB, annual precipitation and runoff were plotted as shown in Figure 4.

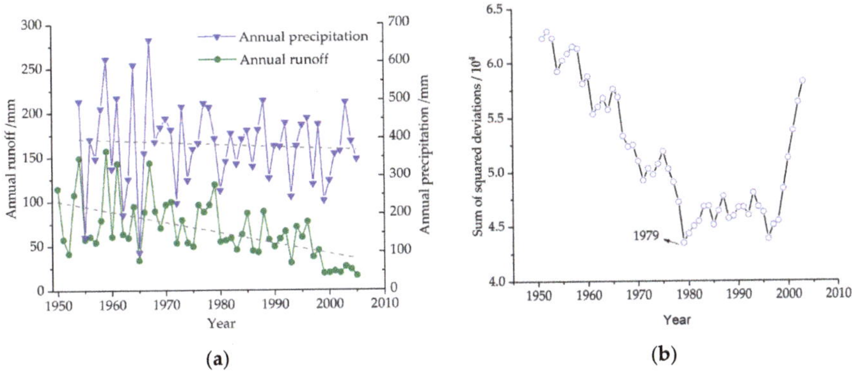

Figure 4. Variation characteristics of precipitation and runoff series. (**a**) The trends in annual runoff and annual precipitation, and (**b**) the result of ordered clustering, showing the variation point.

Figure 4a shows that average annual precipitation in the KRB is close to the horizontal line, indicating that the annual precipitation changed little in the study period. However, annual runoff decreased by more than 50% year by year from 1980 onward.

The results of the OCM are shown in Figure 4b. The figure shows the smallest sum of deviations of squares in 1979, which can be used as a variation point. This result is consistent with the large-scale coal mining that has occurred in the region since 1979 [41].

4.2. ANN Simulation Results

4.2.1. Simulation and Prediction Results

Monthly precipitation from 1954 to 1979 was used as the input and corresponding monthly runoff as the output. The simulated monthly precipitation and monthly runoff results are shown in Figure 5a. Monthly precipitation from 1980 to 2005 was input into the ANN; the monthly runoff predictions for human activities are shown in Figure 5b.

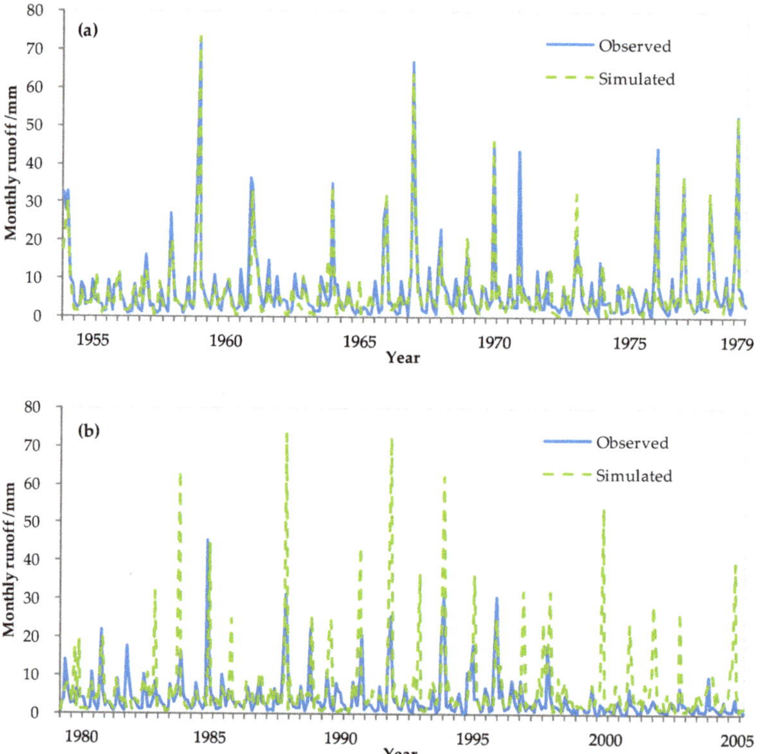

Figure 5. Artificial neural network (ANN) (**a**) simulation results from 1954 to 1979, and (**b**) model prediction results from 1980 to 2005.

In the ANN simulation results in Figure 5a, the observed values are consistent with the simulated values, indicating that simulation quality was satisfactory. The *NSE* value of 0.801 further quantitatively indicates that the ANN model has accurately established the functional relationship between monthly precipitation and monthly runoff.

However, the prediction in Figure 5b shows that the calculated value of the model is much larger than the observed value, indicating a significant error between the calculated value of the model and the observed value. The *NSE* value of −2.312 is far less than 0, which further indicates that the monthly runoff calculated by the model deviates from reality.

4.2.2. ANN Simulation Discussion

The calculation results of the simulation model are relatively high because the ANN model used to establish the relationship between monthly precipitation and monthly runoff is based on historical data, which were not affected by human activities from 1954 to 1979. The model was used to predict the annual runoff from 1980 to 2005, which was a period affected by human activities. However, this deviation reflects the extent of human activities. For example, during the period from 1980 to 2005, the newly built water diversion irrigation system or the introduction of high-water consumption industries, such as coal mining, consumed a large amount of water from the river. The result is that the discharge measured by hydrologic stations was relatively small.

Figure 6 further demonstrates the difference between the simulated monthly runoff and the observed monthly runoff on the annual scale.

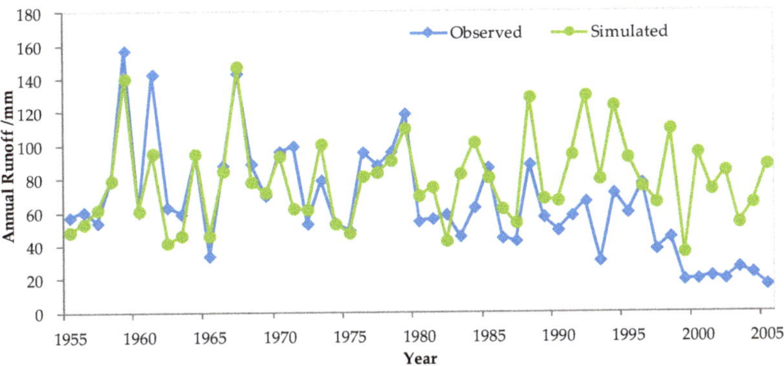

Figure 6. Simulated and observed annual runoff.

Figure 6 shows that in the period from 1954 to 1979, which was not affected by human activities, the simulated results of the model are basically consistent with the observed annual runoff, indicating that the simulated results of the model were accurate. However, during the period from 1980 to 2005, the annual runoff simulated by the model was higher than the observed value. In this period, at the beginning of the 1980s and 1990s, this phenomenon was not obvious. Later, the simulated runoff of the model was seen to be generally larger than the observed runoff. This is directly related to the coal mining that started in the early 1980s [42]. Until the 1980s, almost zero coal was mined annually in the KRB. In the 1980s, the average annual coal mining amount was about 29×10^4 t. In the 1990s, the average annual coal mining amount increased to 520×10^4 t. In the first five years of the 21st century, the average annual coal mining amount increased rapidly to 5452×10^4 t, seriously damaging the ecological environment in the region.

4.3. Results of Drought Risk Assessment

Based on the observed R_1 in the first natural stage, and R_2 and R_{2s} in the second human activity stage, the drought process was identified by the run theory, and the hydrological drought risk in each stage was calculated by the GH Copula function, as shown in Figure 7. The parameters of the Copula function are shown in Table 1.

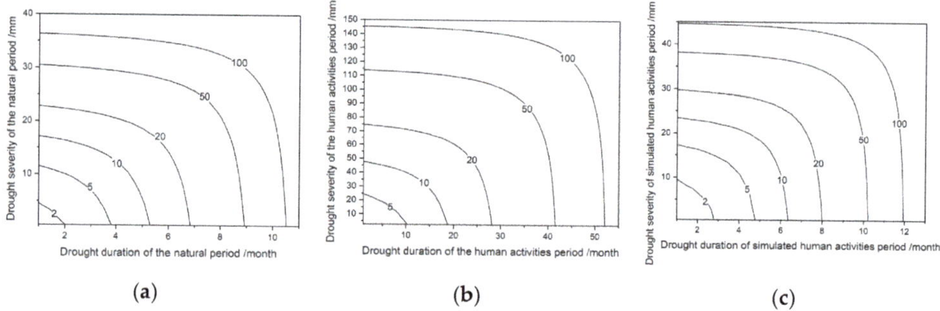

Figure 7. Recurrence period of joint distribution of drought severity and drought duration. (**a**) The risk of natural period, (**b**) for the human activities period, and (**c**) for the simulated human activities.

Table 1. Copula function parameters. R^2—coefficient of determination.

Monthly Runoff Series	Kendall Correlation Coefficient (τ)	Copula Parameters (θ) $\theta = 1/(1 - \tau)$	R^2
1954–1979 observed	0.394	1.650	0.946
1980–2005 observed	0.515	2.154	0.911
1980–2005 simulated	0.438	1.779	0.816

Figure 7 shows that the drought duration and drought severity in the natural state stage in Figure 7a are generally consistent with Figure 7c in the simulated state under each recurrence period. This indicates that without large-scale increase human activities, the drought risk status in the basin from 1980 to 2005 would not have changed significantly.

The results also indicate that climate change has contributed less to the recurrence of drought in the region. However, the drought recurrence period during the stage of human activities (Figure 7b) shows a significant increase in both duration and severity. For example, under the return period of 100 years, the drought duration increased about five times, while the severity of drought increased by nearly four times, severely restricting local social and economic development.

The above results provide a qualitative analysis of the impact of climate change and human activities on drought risk, which needed to be further quantitatively analyzed.

4.4. Quantitative Analysis of Contribution Rate of Climate Change and Human Activities to Drought Risk

According to Equations (7)–(11), the contribution rate of climate change and human activities to the change in drought severity was quantitatively calculated under the condition of drought durations of 1, 2, 5, and 10 months, and the recurrence periods of 2, 5, 10, 20, 50, and 100 years, as shown in Figure 8. The figure shows that human activities and climate change jointly affected the drought severity values under different drought recurrence periods. The red columns are longer, indicating a greater impact of human activity. This is consistent with the little change occurring in annual precipitation (Figure 4a) and large-scale human activity in the area.

Specifically, when the duration of drought is short, one or two months, and the recurrence period is low (two years), the contribution of climate change and human activities to drought severity is opposite compared to under other conditions. Figure 8a–d show that human activities contribute to reducing drought severity in a low drought recurrence period. This shows that local water resources protection measures, such as soil and water conservation, can promote the alleviation of frequent small-scale droughts, though the effect on reducing major drought disasters is limited. The unlimited development of rivers and other water resources, must still be reduced, and corresponding water resources planning and use policies must be formulated.

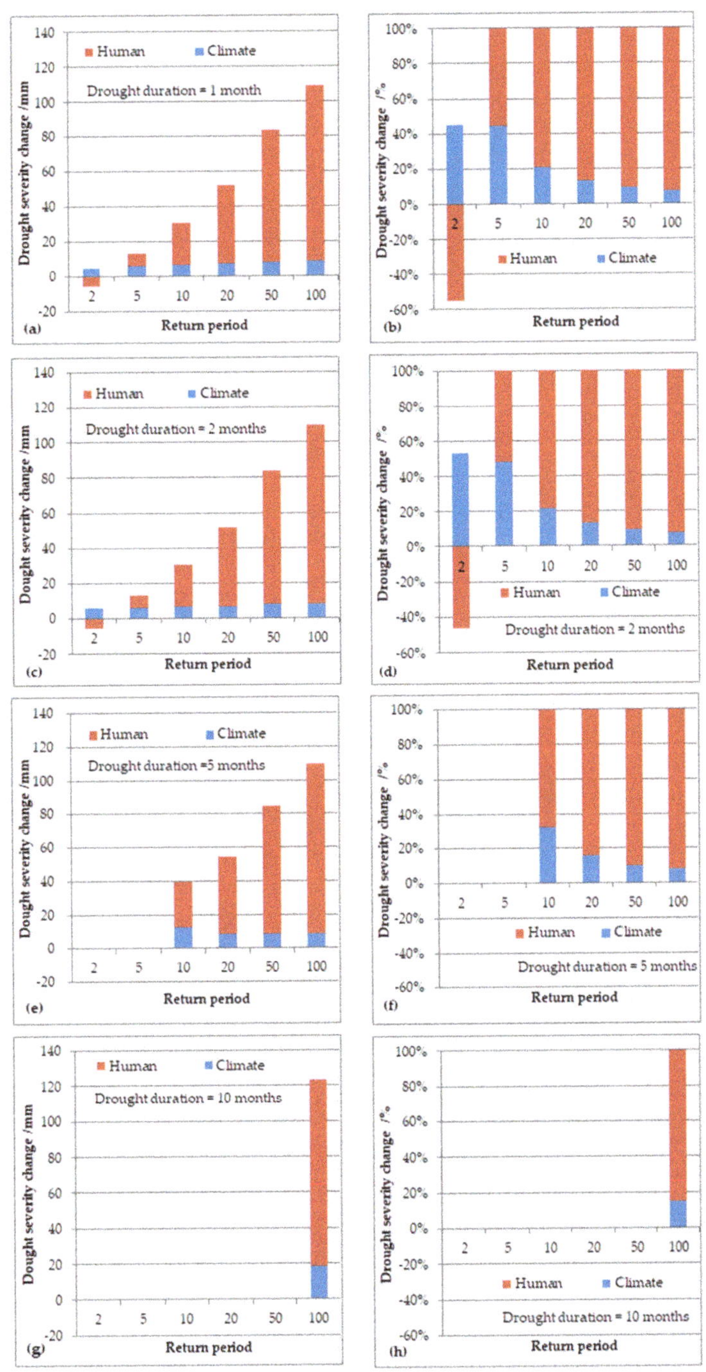

Figure 8. Drought risk attribution contribution: (**a,c,e,g**) Changes in drought severity and (**b,d,f,h**) the corresponding contribution from human activities and climate change.

As the drought duration increased to 5 and 10 months, drought severity in the minor recurrence period could not be calculated. For example, when $D = 5$ months, the drought severity in the recurrence periods of two and five years can no longer be calculated because drought duration and drought severity jointly determine the recurrence period of drought. Greater drought duration is usually accompanied by greater drought severity.

For the 100-year drought recurrence period, the variation in each drought severity value is close to 100 mm. This value is also close to the difference between the simulated annual runoff and the observed annual runoff in Figure 6, likely caused by the excessive exploitation of water resources by local large-scale water industries. Therefore, if this region wants to prevent major drought disaster events that occur once in a century, it is necessary to adjust the current high-consumption water industry situation in the region and realize the sustainable use of water resources [43], for example, through the transfer of high-consumption water industries to other areas with abundant water resources, improving the water use rate of high-consumption water industries or massively increasing the proportion of reclaimed water [44]; the idea of virtual water can also be used as a reference for changing the local grain planting structure [45,46] to reduce the overall risk of large-scale drought disaster.

5. Conclusions

(1) The annual runoff series in the KRB changed significantly after 1979. The annual runoff decreased year over year after the variation, while the average annual precipitation changed little.

(2) ANN can be used to simulate the relationship between monthly precipitation and monthly runoff in the period not affected by human activities, but the simulation results for the period affected by human activities are poor. Human activities became widespread in the area after 1979, leading to a change in the relationship between monthly precipitation and monthly runoff. The difference between simulated and measured results reflects the influence of human activities on runoff and hydrological drought risk.

(3) The quantitative identification of drought risk generally reflects that human activities mainly contributed to the increase in drought risk in this region, but human activities have a beneficial effect on drought events with a short recurrence period of two years and contributed more to high-risk drought events with recurrence of five years and longer.

We mainly focused here on the hazard of hydrological drought risk. Further research should consider the vulnerability and exposure to hydrological drought. Additionally, the methods we used are simple in that the data-driven precipitation–runoff model does not consider hydrological factors such as evapotranspiration, temperature, and soil water storage in the basin. Therefore, further studies should adopt a hydrological model with a physical mechanism.

Author Contributions: Conceptualization and writing—original draft preparation, M.Z.; methodology, M.Z. and R.Z.; validation, R.Z.; data curation, J.W. All the authors read and approved the final manuscript.

Funding: The authors would like to thank the support of the National Key Research and Development Program of China under Grant No. 2017YFC1502405 and No. 2016YFC0401303, the National Natural Science Foundation of China under Grant No. 51579059, 51409001, the Key Research and Development Program of Shandong Province under Grant No. 2017GSF20101, and the Excellent Talents Supporting program of Higher Education of Anhui Province under Grant No. gxyqZD2016127.

Acknowledgments: The authors would like to thank anonymous reviewers for their valuable comments that helped improve the original version of this paper.

Conflicts of Interest: The authors declare no conflict of interest.

References

1. Fleig, A.K.; Tallaksen, L.M.; Hisdal, H.; Demuth, S. A global evaluation of streamflow drought characteristics. *Hydrol. Earth Syst. Sci.* **2006**, *10*, 535–552. [CrossRef]
2. Wilhite, D.A.; Glantz, M.H. Understanding the drought phenomenon: The role of definitions. *Water Int.* **1985**, *10*, 111–120. [CrossRef]

3. Allen, M.R.; Ingram, W.J. Constraints on future changes in climate and the hydrologic cycle. *Nature* **2002**, *419*, 224–232. [CrossRef] [PubMed]

4. Oki, T. Global Hydrological Cycles and World Water Resources. *Science* **2006**, *313*, 1068–1072. [CrossRef] [PubMed]

5. Jin, J.L.; Wei, Y.M.; Zou, L.L.; Liu, L.; Fu, J. Risk evaluation of China's natural disaster systems: An approach based on triangular fuzzy numbers and stochastic simulation. *Nat. Hazards* **2012**, *62*, 129–139. [CrossRef]

6. Song, S.; Singh, V.P. Meta-elliptical copulas for drought frequency analysis of periodic hydrologic data. *Stoch. Environ. Res. Risk Assess.* **2010**, *24*, 425–444. [CrossRef]

7. Mishra, A.K.; Singh, V.P. A review of drought concepts. *J. Hydrol.* **2010**, *391*, 202–216. [CrossRef]

8. Cong, D.; Zhao, S.; Chen, C.; Duan, Z. Characterization of droughts during 2001–2014 based on remote sensing: A case study of Northeast China. *Ecol. Inform.* **2017**, *39*, 56–67. [CrossRef]

9. Yuan, Z.; Yan, D.; Yang, Z.; Yin, J.; Breach, P.; Wang, D. Impacts of climate change on winter wheat water requirement in Haihe River Basin. *Mitig. Adapt. Strat. Glob.* **2016**, *21*, 677–697. [CrossRef]

10. Yuan, X.C.; Wang, Q.; Wang, K.; Wang, B.; Jin, J.L.; Wei, Y.M. China's regional vulnerability to drought and its mitigation strategies under climate change: Data envelopment analysis and analytic hierarchy process integrated approach. *Mitig. Adapt. Strat. Glob.* **2015**, *20*, 341–359. [CrossRef]

11. Dai, A. Increasing drought under global warming in observations and models. *Nat. Clim. Change* **2013**, *3*, 52–58. [CrossRef]

12. Forootan, E.; Khaki, M.; Schumacher, M. Understanding the global hydrological droughts of 2003–2016 and their relationships with teleconnections. *Sci. Total Environ.* **2019**, *650*, 2587–2604. [CrossRef]

13. Xiao, M.Z.; Zhang, Q.; Singh, V.P. Influences of ENSO, NAO, IOD and PDO on seasonal precipitation regimes in the Yangtze River basin, China. *Int. J. Climatol.* **2014**, *35*, 3556–3567. [CrossRef]

14. Strzepek, K.; Yohe, G.; Neumann, J.; Boehlert, B. Characterizing changes in drought risk for the United States from climate change. *Environ. Res. Lett.* **2010**, *5*, 044012. [CrossRef]

15. Kirono, D.G.C.; Kent, D.M.; Hennessy, K.J.; Mpelasoka, F. Characteristics of Australian droughts under enhanced greenhouse conditions: Results from 14 global climate models. *J. Arid Environ.* **2011**, *75*, 566–575. [CrossRef]

16. Wang, L.; Chen, W. A CMIP5 multimodel projection of future temperature, precipitation, and climatological drought in China. *Int. J. Climatol.* **2014**, *34*, 2059–2078. [CrossRef]

17. Yang, X.L.; Zheng, W.F.; Lin, C.Q.; Ren, L.L.; Wang, Y.Q.; Zhang, M.R.; Yuan, F.; Jiang, S.H. Prediction of drought in the Yellow River based on statistical downscale study and SPI. *J. Hohai Univ.* **2017**, *45*, 377–383. (In Chinese)

18. Williams, A.P.; Seager, R.; Abatzoglou, J.T.; Cook, B.I.; Smerdon, J.E.; Cook, E.R. Contribution of anthropogenic warming to California drought during 2012–2014. *Geophys. Res. Lett.* **2015**, *42*, 6819–6828. [CrossRef]

19. Fengqing, J.; Cheng, Z.; Guijin, M.; Ruji, H.; Qingxia, M. Magnification of Flood Disasters and its Relation to Regional Precipitation and Local Human Activities since the 1980s in Xinjiang, Northwestern China. *Nat. Hazards* **2005**, *36*, 307–330. [CrossRef]

20. Al-Faraj, F.A.M.; Scholz, M. Assessment of temporal hydrologic anomalies coupled with drought impact for a trans-boundary river flow regime: The Diyala watershed case study. *J. Hydrol.* **2014**, *517*, 64–73. [CrossRef]

21. Zhang, R.; Chen, X.; Zhang, Z.; Shi, P. Evolution of hydrological drought under the regulation of two reservoirs in the headwater basin of the Huaihe River, China. *Stoch. Environ. Res. Risk Assess.* **2015**, *29*, 487–499. [CrossRef]

22. Luan, Q.H.; Fu, X.R.; Liu, J.H.; Shao, W.W.; Bai, L.L.; Xu, X.Y. Drought events and drought-waterlogging asynchronismin Fen-wei Plain. *South North Water Trans. Water Sci. Technol.* **2016**, *14*, 90–95. (In Chinese)

23. Tu, X.J.; Du, X.X.; Du, Y.L.; Chen, X.H.; Li, K. Multivariate joint design of hydrological drought and impact of water reservoirs. *J. Lake Sci.* **2018**, *30*, 509–518. (In Chinese)

24. Wang, G.; Wang, Y. Managing Water for Sustainable Utilization as China Warms. *Int. J. Environ. Sci. Nat. Resour.* **2019**, *17*, 555952. [CrossRef]

25. Wang, G.; Zhang, J.; Jin, J.; Weinberg, J.; Bao, Z.; Liu, C.; Zhai, R. Impacts of climate change on water resources in the Yellow River basin and identification of global adaptation strategies. *Mitig. Adapt. Strat. Glob.* **2017**, *22*, 67–83. [CrossRef]

26. Xie, P.; Wu, Z.; Sang, Y.F.; Gu, H.; Zhao, Y.; Singh, V.P. Evaluation of the significance of abrupt changes in precipitation and runoff process in China. *J. Hydrol.* **2018**, *560*, 451–460. [CrossRef]

27. Song, X.; Zhang, J.; Zou, X.; Zhang, C.; AghaKouchak, A.; Kong, F. Changes in precipitation extremes in the Beijing metropolitan area during 1960–2012. *Atmos. Res.* **2019**, *222*, 134–153. [CrossRef]

28. Bao, Z.; Zhang, J.; Wang, G.; Fu, G.; He, R.; Yan, X.; Zhang, A. Attribution for decreasing streamflow of the Haihe River basin, northern China: Climate variability or human activities? *J. Hydrol.* **2012**, *460*, 117–129. [CrossRef]

29. Wang, G.; Zhang, J.; Yang, Q. Attribution of Runoff Change for the Xinshui River Catchment on the Loess Plateau of China in a Changing Environment. *Water* **2016**, *8*, 267. [CrossRef]

30. Fahimi, F.; Yaseen, Z.M.; El-Shafie, A. Application of soft computing based hybrid models in hydrological variables modeling: A comprehensive review. *Theor. Appl. Climatol.* **2016**, *128*, 1–29. [CrossRef]

31. Maier, H.R.; Dandy, G.C. Neural networks for the prediction and forecasting of water resources variables: A review of modelling issues and applications. *Environ. Model Softw.* **2000**, *15*, 101–124. [CrossRef]

32. Dawson, C.W.; Wilby, R.L. Hydrological modelling using artificial neural networks. *Prog. Phys. Geogr.* **2001**, *25*, 80–108. [CrossRef]

33. Maier, H.R.; Jain, A.; Dandy, G.C.; Sudheer, K.P. Methods used for the development of neural networks for the prediction of water resource variables in river systems: Current status and future directions. *Environ. Model Softw.* **2010**, *25*, 891–909. [CrossRef]

34. Abrahart, R.J.; Anctil, F.; Coulibaly, P.; Dawson, C.W.; Mount, N.J.; See, L.M.; Wilby, R.L. Two decades of anarchy? Emerging themes and outstanding challenges for neural network river forecasting. *Prog. Phys. Geogr.* **2012**, *36*, 480–513. [CrossRef]

35. Jin, J.L.; Wei, Y.M.; Zou, L.L.; Liu, L.; Zhang, W.W.; Zhou, Y.L. Forewarning of sustainable utilization of regional water resources: A model based on BP neural network and set pair analysis. *Nat. Hazards* **2012**, *62*, 115–127. [CrossRef]

36. Gupta, H.V.; Kling, H. On typical range, sensitivity, and normalization of Mean Squared Error and Nash-Sutcliffe Efficiency type metrics. *Water Resour. Res.* **2011**, *47*, 125–132. [CrossRef]

37. Zhou, Y.L.; Yuan, X.C.; Jin, J.L.; Li, J.Q.; Song, S.B. Regional hydrological drought frequency based on Copulas. *Scientia Geographica Sinica* **2011**, *31*, 1383–1388.

38. Zhang, Q.; Xiao, M.; Singh, V.P.; Chen, X. Copula-based risk evaluation of hydrological droughts in the East River basin, China. *Stoch. Environ. Res. Risk Assess.* **2013**, *27*, 1397–1406. [CrossRef]

39. Shiau, J.T. Return period of bivariate distributed hydrological events. *Stoch. Environ. Res. Risk Assess.* **2003**, *17*, 42–57. [CrossRef]

40. Wang, G.Q.; Zhang, J.Y.; Xuan, Y.Q.; Liu, J.F.; Jin, J.L.; Bao, Z.X.; Yan, X.L. Simulating the Impact of Climate Change on Runoff in a Typical River Catchment of the Loess Plateau, China. *J. Hydrometeorol.* **2013**, *14*, 1553–1561. [CrossRef]

41. Guo, Q.L.; Yang, Y.; Xiong, X. Using hydrologic simulation to identify contributions of climate change and human activity to runoff changes in the Kuye river basin, China. *Environ. Earth Sci.* **2016**, *75*, 417–426. [CrossRef]

42. Jiang, X.H.; Gu, X.W.; He, H.M. The influence of coal mining on water resources in the Kuye river basin. *J. Nat. Resour.* **2010**, *25*, 300–307.

43. Zhang, M.; Zhou, J.H.; Zhou, R.J. Interval Multi-Attribute Decision of Watershed Ecological Compensation Schemes Based on Projection Pursuit Cluster. *Water* **2018**, *10*, 1280. [CrossRef]

44. Zhou, R.J.; Wang, Y.B.; Zhang, M.; Yu, P.X.; Li, J.Y. Adsorptive removal of phosphate from aqueous solutions by thermally modified copper tailings. *Environ. Monit. Assess.* **2019**, *191*, 198–210. [CrossRef]

45. Wang, Z.Z.; Zhang, L.L.; Zhang, Q.; Wei, Y.M.; Wang, J.W.; Ding, X.L.; Mi, Z.F. Optimization of virtual water flow via grain trade within China. *Ecol. Indic.* **2019**, *97*, 25–34. [CrossRef]

46. Wang, Z.Z.; Zhang, L.L.; Ding, X.L.; Mi, Z.F. Virtual water flow pattern of grain trade and its benefits in China. *J. Clean. Prod.* **2019**, *223*, 445–455. [CrossRef]